Studies in Computational Intelligence

Volume 852

Series Editor

Janusz Kacprzyk, Polish Academy of Sciences, Warsaw, Poland

The series "Studies in Computational Intelligence" (SCI) publishes new developments and advances in the various areas of computational intelligence—quickly and with a high quality. The intent is to cover the theory, applications, and design methods of computational intelligence, as embedded in the fields of engineering, computer science, physics and life sciences, as well as the methodologies behind them. The series contains monographs, lecture notes and edited volumes in computational intelligence spanning the areas of neural networks, connectionist systems, genetic algorithms, evolutionary computation, artificial intelligence, cellular automata, self-organizing systems, soft computing, fuzzy systems, and hybrid intelligent systems. Of particular value to both the contributors and the readership are the short publication timeframe and the world-wide distribution, which enable both wide and rapid dissemination of research output.

The books of this series are submitted to indexing to Web of Science, EI-Compendex, DBLP, SCOPUS, Google Scholar and Springerlink.

More information about this series at http://www.springer.com/series/7092

Szczepan Paszkiel

Analysis and Classification of EEG Signals for Brain–Computer Interfaces

 Springer

Szczepan Paszkiel
Department of Biomedical Engineering,
Faculty of Electrical Engineering, Automatic
Control and Informatics
Opole University of Technology
Opole, Poland

ISSN 1860-949X   ISSN 1860-9503   (electronic)
Studies in Computational Intelligence
ISBN 978-3-030-30583-3   ISBN 978-3-030-30581-9   (eBook)
https://doi.org/10.1007/978-3-030-30581-9

This Springer imprint is published by the registered company Springer Nature Switzerland AG
The registered company address is: Gewerbestrasse 11, 6330 Cham, Switzerland

Contents

Chapter 1
Introduction

This monograph is a collection of information on the development of brain–computer (BCI) technology with particular focus on data acquisition methods-tools used for human brain activity. Due to universal and easy application, author focused on the use of electroencephalography as an essential method commonly used in measurements for the needs of the development of BCI technology. This method makes it possible to investigate bioelectric activity of neurons by placing electrodes directly on the cortical surface (invasive method) or on the head surface (non-invasive method). The signal received during such implementations is an electroencephalographic signal, in which brain waves oscillations such as: alpha, beta, theta, gamma, lambda band oscillations etc., are separated. However, electroencephalography is not the only method of acquisition used during the realization of solutions within the brain—computer interface technology, therefore, the methods of human brain investigations such as: Magnetoencephalography (MEG), Functional Magnetic Resonance Imaging (FMRI), Positron Emission Tomography (PET), Near Infrared Spectroscopy (NIRS) are discussed in this monograph. The methods of data analysis in the scope of human brain activity, including, among others, statistical methods are described in the following chapters. Also, the Moore-Penrose pseudoinverse as a potential tool for the EEG signal reconstruction is presented. Furthermore, the use of the LORETA method for localization of the EEG signal sources in BCI technology is discussed; it is the method for brain activity imaging based on electroencelographic and magnetoencephalographic records. The monograph also discusses the issue of using neural networks for classification of the changes in the EEG signal based on facial expressions, which was then implemented in practical implementations of the developments based on the research results.

The implementation part of the monograph refers to the author's use of BCI technology in control processes. An idea of controlling a mobile vehicle based on facial expressions, which generating an artifact in the EEG signal adequate to the performance of a given activity, was classified for the needs of the process of controlling a mobile robot. Another example of practical implementation refers to the original

© Springer Nature Switzerland AG 2020

S. Paszkiel, *Analysis and Classification of EEG Signals for Brain–Computer Interfaces*, Studies in Computational Intelligence 852,
https://doi.org/10.1007/978-3-030-30581-9_1

use, in the scope of realization methods of control in BCI technology, of LabVIEW environment.

A dynamically developing Virtual Reality (VR) and Augmented Reality (AR) technology has become an impetus for developing the concept of combining AR with BCI technology and then the application of VR technology in correlation with BCI technology. Within the research work, also an exemplary video game in UNITY environment was developed which may be successfully used in a widely developed neurogaming basing on BCI technology, which is described in one of the subsequent chapters.

Within the developments being the outcome of the research work on the brain–computer technology, including identification of the sources of the brain signals generation due to correlation of neuronal cell fractions [1], the possibility of implementation of the solutions coming from BCI technology in the scope of the popular IoT technology in the aspect of smart homes is also presented.

The monograph ends with a chapter summing up the obtained results of the research works, with particular focus on their application possibilities in the aspect of carrying out developments in brain–computer technology.

Reference

1. Accardo, A., Affinito, M., Carrozzi, M., Bouquet, F.: Use of the fractal dimension for the analysis of electroencephalographic time series. Biol. Cybern. **77**, 339–350 (1997)

Chapter 2
Data Acquisition Methods for Human Brain Activity

2.1 Electroencephalography

Clinical electroencephalography is one of several methods of data acquisition from human brain. It was introduced by Hans Berger, a German psychiatrist in the 1930s [3]. It is a noninvasive method consisting in detection and registration of electrical activity of the brain using electrodes attached to the scalp which register changes of electric potential on the skin surface coming from the activity of cerebral neurons [2] and after their amplification they form a record—an encephalogram. The value of the potential registered by consecutive electrodes can be described by Eq. (2.1).

$$V_n = V_{EEGn} + V_{CMS} \qquad (2.1)$$

where: V_n—potential value on electrodes, V_{EEGn}—potential connected with electrical activity of the brain. V_{CMS}—common signal on all electrodes, also connected with interference from the network.

Presently, EEG is most often used by neurologists to differentiate functional from organic brain diseases, to diagnose sleep disorders, headaches, dizziness, to monitor brain activity during heart operations. EEG offers high temporal resolution which is not possible with MRI. Obtaining electrode resistance not exceeding 10k Ω at the start of the investigation is an essential condition for obtaining a good quality EEG. It depends, to a large extent, on proper preparation of the scalp, which before the electrodes are attached, should be carefully degreased and the superficial calloused layer of the epidermis removed.

The disadvantage of using an electroencephalograph in practice is, among others, limitation of resolution by equipment capabilities and the need to use a computer to view and analyze data.

© Springer Nature Switzerland AG 2020
S. Paszkiel, *Analysis and Classification of EEG Signals for Brain–Computer Interfaces*, Studies in Computational Intelligence 852,
https://doi.org/10.1007/978-3-030-30581-9_2

2.1.1 EEG Signal

The EEG signal is a recording of the electrical activity of the brain which measures field potential in the space around neurons. The more synchronized is a given population of neurons, the easier it is to measure the EEG signal. This signal does not have constant amplitude and frequency, the waveform is never a simple harmonic signal. The EEG signal is characteristic of a high time activity of about 1 ms [5].

In this signal we can separate a few types of activities which are characteristic of specific signal frequencies and amplitudes such as: Alpha, Beta, Delta, Gamma, Theta, Mu. Research showed that both the frequency of brain waves and their amplitude are not strictly constant and strictly depends on the activity the brain performs. Pyramidal cells of the cerebral cortex, due to their characteristic location in the cerebral cortex structure are considered to be the main source of the electroencelographic signal in human brain [4, 6].

Alpha waves occur in the frequency range of 8–12 Hz and are characterized by a multiname amplitude of from 3 to about 50 μV on average. The activity of these waves increases when we lay with eyes closed before falling asleep. They have smaller frequency and higher voltage. Alpha activity is present when a person is at rest, in a relaxed state, meditating etc. These waves disappear during increased mental effort, for example, when a person tested performs complex mathematical tasks. The research carried out shows that that the suppression of alpha rhythm is the result of biological activity desynchronization process due to, for example, sensory concentration. One of the types of activity are Beta waves with a frequency range of between 12 and about 28 Hz and their characteristic feature is a small signal amplitude. The electroencelographic signal with Beta waves activity shows the features of desynchronization. Beta waves can be split into: SMR—the sensorimotor rhythm in the band in the range of 13–15 Hz; Low Beta Waves—is the rhythm corresponding to a proper dynamics of excitement processes, dominates in the left brain hemisphere; High Beta Waves—the band above 18 Hz, associated with hyperactivity of nervous processes. It has been verified that Beta waves characterize normal waking consciousness, consisting in receiving stimuli from the environment involving the following senses: hearing, vision, taste, touch and smell. Anger, fear, surprise are a few of the states that are characteristic for the states occurring during Beta waves activity. The amplitude of Beta waves usually does not exceed 20 μV. Theta waves are the waves with a frequency range of 4–7 Hz which can be observed especially during intensive emotions. They are characteristic for the processes of deep meditation. Theta waves can be divided into two levels: Low Theta—activity rhythm in the band from 4 to 5.5 Hz and High Theta—rhythm in the range from 5.5 to 7 Hz. This rhythm is connected directly with electrical activity of the hippocampus and its signal amplitude is high, of the value up to 100 μV. The lowest observed Theta wave value is about 30 μV. If Theta waves are registered only in one location or if they dominate one hemisphere, it is very likely that it indicates structural damage. Theta wave activity changes as we age. Delta waves with a frequency range of 1–3 are of high signal amplitude in the range from 100 to 200 μV. Their highest activity occurs during deep sleep. Delta waves, during the sleep of the person examined, alternate with discharges consisting of Alpha waves. The characteristic feature of these waves is that they do not occur in a normal EEG recording of an adult who is awake. Their presence always indicates a brain disfunction. They occur only in older people but in small numbers, especially in the temporal areas. Gamma waves are observed in the range from 30 to 200 Hz. Studying literature carefully we can also find waves in the range up to 500 Hz. Te amplitude of these waves is within 3–55 μV. Numerous investigations show that the amplitude of these waves increases if the person examined concentrates on the stimulus source. Gamma wave duration time is very short and it is contained within several dozen ms. Mu waves, also called wicket rhythms, are a physiological phenomenon. They are registered from central leads, from over the motor cortex. They are well defined, appear in bursts of 7–11 Hz and resemble the Greek letter *Mi*. Mu activity sometimes shows more clearly during sleepiness and recording with open eyes. Mu diminishes with upper limb movements, e.g. during clenching the fists or even thinking about this action. It is often very clear on the craniotomy side. The meaning of this rhythm consists mainly in its recognition as a normal rhythm. Lambda waves are electropositive potentials registered transitorily in the occipital regions. They are sharp, usually symmetrical and can be mistaken for epileptic potentials. They are evoked by visual scanning of a picture. As an example may serve the work: *The Knights of the Round Table* which was used for the first time by *Bickford*—one of the most respected electroencephalographist *in Mayo Clinic*, a non-profit organization, which is a medical research group in the United States. A subject instinctively explores interesting items and Lambda waves are clearly elicited. Probably they are visual potentials evoked [1]. The main advantage from getting to know Lambda waves is the knowledge that they are a physiological phenomenon and not epileptiform discharges.

2.1.2 Artifacts in EEG Signal

Artifacts occur in the EEG signal if it is not proportional to the eclectic potential generated by the brain. They distort an actual course of brain waves. Artifacts can be classified depending on their origin as technical and biological artifacts. The influence of artifact sources is directly proportional to the amplitude of the signal generated and inversely proportional to the distance between the sources and EEG electrodes. Also the changes in conductivity between the electrodes and the brain should be recognized as artifacts.

Technical artifacts are removed from the EEG signal by using a low-pass filter. The sources of technical artifacts, among others, are: electricity supply, medical apparatus and workstations etc.

Biological artifacts originate from: eyeball movements, skeleton muscle movements, body movements in relation to electrodes (head tremor), heart beat, arterial pulsation, perspiration, secretion of sebum on the skin, teeth clenching, swallowing. Generally, the source of biological artifacts are all organs in an organism excluding the brain and tissue. It is inferred that the eyeball movement will cause inferences of the highest amplitude, i.e. inferences that obscure to the largest extent the image of brain waves. Moreover, in the electroencephalographic signal artifacts such as impulses caused by muscle stimulation occur, the spectrum of which overlaps the spectrum of the very brain activity and thus makes their removal difficult. Cardiac inferences may be removed by selecting proper reference electrodes which make it possible to measure the difference of potentials within the brain. Psychological condition of the person examined also influences directly the quality of the EEG signal, including biological inferences that may occur in this signal. An objective assessment of the influence of each interference source on the EEG signal recording should be a significant observation towards the elimination of artifacts.

The elimination of artifacts may be carried out by liquidating the source of interference from which they originate. The more precisely the source of interference can be characterized, the better it can be eliminated from measurement. If it is impossible to separate the source of interference or to cut off the channel of interference propagation, the only thing that can be done is processing of the recorded encephalogram to recover undistorted EEG. Bandpass filtration is the easiest and commonly used method of interference elimination. This filtration consists in suppressing all frequencies except for the band of a useful signal. The interferences of the frequency band overlapping the brainwave band cause problems. Electromagnetic field generated by 50 Hz powerlines can serve as an example. A commonly used narrowband filter cuts out from the recoding a 50 Hz component, but both a powerline signal and the signal generated by the brain become suppressed. When it is impossible to filter out interferences and obtain an undistorted record of brainwaves, it is necessary to exclude from investigation all segments of the recording in which the occurrence of artifacts is suspected.

Detecting artifacts of a relatively big amplitude compared with the EEG record is not difficult using either a visual method or automatic analysis. There exist analysis algorithms for detecting specific kinds of artifacts, e.g. coming from the eyeball movements. Such searches result in elimination of the segments with interferences and when the nature of the inferences is well known, it is possible to recover a proper record. Finding interferences of relatively small amplitude values is more difficult. It is possible to claim the occurrence of artifacts only based on the spectral or topographic analysis. As it results from the analyses carried out, for example, the eyeball movement has the biggest influence on the potentials of frontal electrodes, the skeleton muscle movement manifests itself in the form of interferences on lateral leads. Artifacts caused by displacement of an electrode in relations to a body cause interferences only in one own registration channel, therefore they are quite easy to observe. The elimination of external artifacts is not a big problem. It is much more difficult to eliminate biological (physiological and muscle) artifacts.

2.2 Magnetoencelography

Magnetoencelography (MEG) deals with registration of the magnetic field generated by the brain. It makes it possible to obtain a signal of a much higher spatial resolution than EEG and also an analysis in a much wider frequency range. The MEG signal is much more sensitive to the activity of the neuron population parallel than perpendicular to the scalp. According to Maxwell equations, each electric current generates perpendicularly oriented magnetic field and it is the field strength that is measured by sensors. Difficulties with measuring brain magnetic fields are connected with separating the fields coming from the brain from electromagnetic noise. Magnetic fields generated by a metal screwdriver movement in the distance of a few meters away or a passing car several dozen meters away are several orders of magnitude stronger than the fields coming from the brain. Therefore the measurement of MEG must be taken in specially designed shielded rooms.

However, the information carried by the energy changes of this signal is adequate to the information we infer from ERD/ERD waveforms of the EEG signal. MEG is used in scientific research the aim of which is to determine the functions of the particular brain areas; in clinical diagnostics and as an investigation performed during neurosurgical operations to localize pathological regions. Furthermore, it is also useful in the method called neurofeedback.

2.3 Functional Magnetic Resonance Imaging

Functional magnetic resonance imaging (FMRI) is the third specialized variety of imaging using magnetic resonance which measures an increase of blood flow and oxygenation of the active part of the brain [7]. This method takes advantage of the fact that during nervous cell activity their oxygen demand increases and carbon dioxide production intensifies. An increase of a given region activity is measured with BOLD—Blood-oxygenation-level-dependent—response, which defines the dependence of magnetic resonance signal intensity on the level of blood oxygenation. The concept of FMRI is based on the use of MRI investigation and extending it by observation based on the properties of oxygenated and deoxygenated blood. The object under study is placed in a strong magnetic field of parallel field lines. Scanner coils send with a definite frequency short-duration electromagnetic pulses towards the object under study causing excitation of proton spins in the hydrogen atomic nuclei, which are components of water molecules, which is found in living organisms. For a constant magnetic field of 1.5 T, this frequency is of about 63 MHz. The pulse makes the atom nuclei magnetized and they become themselves the source of the electromagnetic field. After the pulse stops acting, electromagnetic radiation caused by the return of the spins to non-excited state is registered by coils which act as receivers. Protons returning to their ground state emit an electromagnetic wave weakening over time of a similar frequency as the frequency from which an electromagnetic pulse was sent in their direction. The speed of this wave attenuation depends on characteristic magnetic properties of the atoms of particular tissues. The registration of these waves using the so-called constant magnetic field gradients, makes it possible to reproduce the image of the interior of the object under study using a computer.

2.4   Positron Emission Tomography

Positron Emission Tomography (PET) is an imaging technique in which, unlike in computed tomography, instead of external source of X-ray or radioactive radiation, the radiation emitted during positron annihilation is registered. The source of positrons is a radioactive tracer introduced into a patients body which decays as Beta plus. This tracer contains radioactive isotopes of a short half-life therefore most radiation is formed during an investigation, which limits tissue damage caused by radiation. It is also connected with the need to launch a cyclotron nearby, which considerably increases the costs.

Presently, practically all available positron emission tomography scanners are hybrid devices of the type: PET-CT, PET/CT—PET combined with a multirow computed tomography scanners, PET-MRI, PET/MRI—combination of PET with magnetic resonance. The positrons which occur due to radioactive decay, after covering the distance of a few millimeters, collide with electrons in the tissue and become annihilated. Due to annihilation of the electron-positron pair, two quanta of electromagnetic radiation are formed (photons) of the energy value of 511 keV each, moving in opposite directions (at a 180° angle). These photons are registered simultaneously by two out of the many detectors set at various angles in relation to the patient's body (usually in the form of a ring) and thus an exact place of positron occurrence can be determined. This information is registered in a digital form on a computer disc, which makes it possible to construct images that are cross sections of the patient's body, analogous to the images obtained by magnetic resonance imaging (a 3-dimensional image of the object investigated is obtained).

2.5   Near Infrared Spectroscopy

Near Infrared Spectroscopy (NIRS) is another visualization technique of brain activity, consisting in passing laser beams through the scull. These lasers are very weak, however they use the light wavelengths close to infrared from about 700 to 2500 nm, for which the scull is transparent. Blood with oxygen absorbs other light waves frequencies than the blood in which oxygen has been absorbed. Therefore, observing the amount of light of various frequencies reflected from the brain, researchers may watch the blood flow. In the case of creating the activation map, this technique is called diffuse optical tomography DOT. In the case of registering light diffusion due to the changes in cells occurring during the excitement of neurons, it is an event-related optical signal—EROS. EROS is a technique of brain scanning which uses infrared light through optic fibres to measure the changes of optical properties of the activated regions of the cerebral cortex. While the techniques such as diffuse optical tomography and NIRS measure optical absorption of hemoglobin, and thus are based on blood flow, EROS takes advantage of the scattering properties of the neurons themselves, and thus provides a much more direct measure of cellular activity. EROS can pinpoint activity in the brain within millimeters and milliseconds. Currently, its biggest limitation is the inability to detect activity more than a few centimeters deep, which limits fast optical imaging to the cerebral cortex.

References

1. Badcock, N.A. et al.: Validation of the Emotiv EPOC EEG system for research quality auditory event-related potentials in children. Peer J. 907 (2015). http://dx.doi.org/10.7717/peerj.907
2. Bear, M.F., Connors, B.W., Paradiso, M.A.: Neuroscience: Exploring the Brain. Lippincott, Philadelphia (2006)
3. Berger, H.: Uber das Elektrenkephalogramm des Menschen. Archiv. Psychiatrie. und. Nervenkrankheiten. **87**, 527–570 (1929)
4. Ovaysikia, S., Tahir, K.A., Chan, J.L., DeSouza, J.F.X.: Wordwinsoverface: Emotional stroop effect activates the frontal cortical network. Front. Hum. Neurosci. (2011). https://doi.org/10.3389/fnhum.2010.00234
5. Paszkiel, S.: Concept of expert system interpreting correctness of measurement and method of the EEG signal analysis for needs of the brain-computer interface. Poznań Univ. Technol. Acad. J. Electr. Eng. **88**, 209–216 (2016). ISSN 1897-0737
6. Paszkiel, S.: The population modeling of neuronal cell fractions for the use of controlling a mobile robot. In: Measurements Robotics Automation—P A R, 2/2013, Warszawa, pp. 254–259 (2013)
7. Valente, G., Kaas, A.L., Formisano, E., Goebel, R.: Optimizing FMRI experimental design for MVPA-based BCI control: combining the strengths of block and event-related designs. NeuroImage **186**, 369–381 (2019). https://doi.org/10.1016/j.neuroimage.2018.10.080

Chapter 3
Brain–Computer Interface Technology

BCI technology dates back to the 1990s and even though it is relatively new, it may significantly revolutionize in the future the solutions which people use to communicate with computers and other devices [6]. All presently widely-used methods that a user may give a computer a kind of command are based on muscle movement. Both a traditional mouse and a touchpad or a touch screen or even voice or eyeball control require muscle movement. This movement is not generated instantly as it is a reaction of a human organism to a signal coming from the brain and therefore it is burdened with additional delay and energy input connected with muscle movement. BCI interfaces (Fig. 3.1) avoid this unnecessary step on the information path from the brain to the computer through the acquisition and analysis of the signals directly from human brain. Therefore, in theory it is a solution much faster and accurate than all other interfaces of a user's communication with the computer.

Each interdependence in neuronal structures which is the result of correlation between neuronal cell fractions is connected with the activity causing changes in electrical activity of the brain. Reading this activities and converting them to the data that can be understood by a computer is the main task of the interfaces based on BCI technology [2]. There are distinguished two data acquisition methods: invasive method based on electrodes implanted directly to the cerebral cortex and non-invasive method acquiring data on brain activity only from the level of the sculp. The invasive method provides more accurate readings of the bioelectrical activity of brain than the non-invasive method as the electrodes in the cerebral cortex can detect even small changes of potential, avoiding additional biological barriers such as the skull and the scalp. This method, however, is not used on a wide scale due to the need of interference in human body and possible negative effects observed during its first implementations. After extended use, the electrodes in the brain may ulcerate and loose their properties and their replacement becomes necessary. With each surgery there also appears a risk of infection dangerous to the user's health. Therefore, this

© Springer Nature Switzerland AG 2020
S. Paszkiel, *Analysis and Classification of EEG Signals for Brain–Computer Interfaces*, Studies in Computational Intelligence 852,
https://doi.org/10.1007/978-3-030-30581-9_3

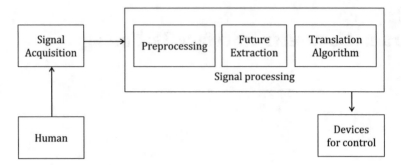

Fig. 3.1 General structure of BCI system

method is popular mainly among paralyzed people and until now, its implementation took place only in definitely justified cases.

The non-invasive method is much more popular owing to a more and more wide range of devices available for commercial use. It is less complicated than the invasive implementation and does not require any interference in human organism. Furthermore, a device purchased may be used by many users as it operates with reusable electrodes placed on the scalp. In most cases these electrodes require for their proper operation only soaking with physiological saline or applying on them a thermal paste to change the character of ion conductivity by which a signal is transferred in a human body to electron conductivity in a device. Soaking the electrodes in the case of physiological saline also allows to decrease resistance of these electrodes and thus improve the parameters of signal reception. Many factors depending on a user and his environment have influence on the parameters deciding on the quality and processing time of the EEG signal read in the case of BCI technology based on electroencephalography [5]. Electromagnetic signals identified from the user's environment constitute an obvious obstacle in reading the electrical signal from the brain which is characterized only by small changes of electrical potential on the skin. Biological factors such as, among others: high blood pressure, higher pulse or excessive stress are further artifacts impeding full use of devices based on BCI technology.

Non-invasive methods of processing signals from the user's brain are most often based on two methods of data acquisition: EEG and MEG differing first of all in the type of the information read—electrical potential on the scalp in the case of EEG electroencephalography and the magnetic field, inseparably connected with brain activity in magnetoencephalographic investigation MEG. In commercial appliances, electroencephalography is used more often due to the possibility of obtaining high resolution measurements, short delays and lower implementation costs.

From the physical point of view, the brain is an electrical appliance generating, during the process of thinking, electrically changing currents in the areas of both frequency and amplitudes. Based on Maxwell theory, the changes in the magnetic electric field are accompanied with the occurrence of the electromagnetic field of the same frequency. This phenomenon is presented with equation:

$$\nabla^2 E = u_0 e_0 \frac{\partial^2}{\partial t^2} E \tag{3.1}$$

$$\nabla^2 B = u_0 e_0 \frac{\partial^2}{\partial t^2} B \tag{3.2}$$

where: E—electric field strength; B—magnetic induction; u_0—vacuum permittivity; e_0—permeability. Based on the equation describing physical phenomena taking place in human brain it can be stated that the brain is a transmitter generating the electric and magnetic field operating both during man's activity and during sleep. In order to receive information coming from the brain the so-called electroencelograph is used.

BCI interfaces constitute a useful solution for people with irreversible movement disorders or visual system disorders. Paralyzed people. in spite of their disease can still express their will to move some muscle group and this will is reflected in the form of an electric pulse in the brain. Reading this pulse and transforming it into a proper movement of mechanical devices such as various types of artificial limbs and wheelchairs for the disabled are already used in bringing back the possibility of normal functioning to sick, disabled people [3]. The development of BCI interfaces, increasing the possibility and quality of reading these pulses brings hope for a more perfect equipment which will enable these people the best possible functioning in spite of their disease. Also the application of the BCI interface for blind people brings promising results. It was back in 1999 that the researchers from California managed to decode the image seen by a cat using the BCI interface. Further research in this scope also made it possible to reconstruct a video material based on the brain activity evoked by watching scenes by a patient. Therefore it could be deduced that a reverse situation, in which proper brain stimulation will make it possible to generate a signal that will stimulate a signal formed using the organs of vision, is possible. As a consequence, in 2002 an ocular prosthesis was implanted in the patient suffering with acquired blindness, which enabled this person to see the world around him in a basic way.

The development of BCI is becoming one of the most innovative paths of development of modern automation, robotics and informatics [4]. BCI technology studied so far by many research institutes in the world has a chance to become one of the means of communication with surrounding equipment, not only with a computer. It is worth mentioning that presently also numerous private investors contributing to the progress in this field are involved in the development of technologies connected with BCI. One of the very futuristic initiatives of this kind is the Neuralink project developed by Elon Musk the aim of which is to improve human brain based on BCI. Apart from the already known possibility of controlling computers using information read from the brain, the Neuralink project is to "open human brains" to the possibilities of bilateral flow of data between the brain and informatics systems and even between various brains. In the future this functuality may allow for some kind

Fig. 3.2 NeuroSky
MindWave Mobile 2 from
Neurosky

brain actualization with required data or skills, or for direct distance communication resembling telepathy.

One of the manufacturers of BCI appliances is Neurosky, Inc founded in the Silicon Valley in California in 2004. It derived its name from neurosky technology which was developed five years earlier in 1999. The company creates appliances using the brain–computer interface technology adapting electroencephalography (EEG) and electromyography (EMG) non-invasive technology consumer product applications. MindWave Mobile 2, which appeared on the market in June 2018, is the latest Neurosky product (Fig. 3.2).

NeuroSky MindWave Mobile 2 allows for realization of many possibilities. It can be used, among others, for: game service—better game experience; education—applications making it possible to visualize how human brain works; cursory verification of health condition—applications for checking the mood and for meditation; carrying out research—the possibility of new inventions in the scope of EEG products and neuroscience.

Emotiv Inc., which was founded in 2013, is another leading brand in BCI technology. Earlier, in 2003 it existed under the firm Emotiv Systems, and later in 2010, under the firm Emotiv Lifesciences Inc. It is known as a manufacturer of the products using brain–computer interface technology and biotechnology. The vast majority of the products released by this company is worn on the body and use a non-invasive EEG technology. The Emotive company products are as follows: headsets; branded Software Development Kits (SDK) for programmers [1]; software; mobile applications. Figures 3.3, 3.4 and 3.5 show currently (2019) available appliances offered by

Fig. 3.3   Emotiv EPOC+ NeuroHeadset from Emotiv Inc.

Fig. 3.4   Emotiv Insight from Emotiv Inc.

Emotive, such as: Emotiv EPOC+ NeuroHeadset, Emotiv Insight and Emotiv EPOC Flex.

Emotiv EPOC+ NeuroHeadset is designed for scalable and contextual human brain research. It provides a professional access the data read out from brain owing to a fast and easy operation of the device. Its big advantage are 9 axis motion sensors, a battery and a wireless connection between a headset and another device. Emotiv EPOC+ NeuroHeadset has: 14 channel sensors, 2 referencial sensors, wireless Bluetooth connection, a battery, a USB port.

Fig. 3.5 Emotiv EPOC Flex
from Emotiv Inc.

Emotiv Insight is a simpler than Emotiv EPOC+ NeuroHeadset device, designed for everyday use. Advanced electronics and full optimization ensure clean, robust EEG signals anytime, anywhere.

Emotiv EPOC Flex has 32 measuring sensors and is very accurate as far as signal reading out is concerned. It has wireless technology, is elastic and adjusts to the head shape.

Emotiv Inc. makes the platform with the tools required for creating, managing and scaling applications and projects, called Cortex, available to programmers. Cortex facilities creating games and applications and storing data from experiments. Emotiv, Inc. gives access to its proprietary software supporting work with its equipment. Emotiv software includes: EmotivPRO—integrated software solution for neuroscience research and education; EmotivBCI—flagship software for brain computer interface; BrainViz—real time 3D brain visualization software. MyEmotiv is a mobile application available on Google Play and in AppStore. It is for monitoring brain activity during daily life, collecting data and therefore improving mental performance and well-being of a user.

References

1. Emotiv Software Development Kit User Manual for Release 2.0.0.20
2. Kaur, B., Singh, D., Roy, P.P.: EEG Based emotion classification mechanism in BCI. Procedia Comput. Sci. **132**, 752–758 (2018). https://doi.org/10.1016/j.procs.2018.05.087
3. Kübler, A., Birbaumer, N.: Brain-computer interfaces and communication in paralysis: extinction of goal directed thinking in completely paralysed patients. Clin. Neurophysiol. **119**(11) (2008)

4. Paszkiel, S.: The use of brain computer interfaces in the control processes based on industrial PC in terms of the methods of EEG signal analysis. J. Med. Inform. Technol. **22**, 55–62 (2013). ISSN 1642-60372013
5. Qin, Z., Li, Q.: High rate BCI with portable devices based on EEG. Smart Health **9–10**, 115–128 (2018). https://doi.org/10.1016/j.smhl.2018.07.006
6. Wolpaw, J.R., Birbaumer, N., McFarland, D.J., Pfurtscheller, G., Vaughan, T.M.: Brain-computer interfaces for communication and control. Clin. Neurophysiol. **113**(6), 767–791 (2002)

Chapter 4
Using the Moore-Penrose Pseudoinverse for the EEG Signal Reconstruction

The EEG signal which is obtained as a result of acquisition with the electrodes placed of the scalp of the person examined is subjected to the processes of verification and classification [2]. It is also often important to determine the source of a signal in human brain and thus separate interference. There are a lot of techniques of elimination of such artifacts including, among others, Blind Signal Separation (BSS), which carries out separation of unknown signals with lack of information on the way of their mixing [6]. The fundamental assumption of the Moore-Penrose pseudoinverse application in the EEG signal reconstruction is the possibility of identifying the location from which a given signal is coming [3]. Also it is worth keeping in mind that in practice the number of brain signal sources is bigger than the number of measuring electrodes. During electroencephalographic measurements we observe the potential occurring on the scalp based on local activities of neuronal cell fractions. Many scientists in their research aim at estimating the location of the source/sources of this potential at a given moment of time. Hence in practice the cerebral cortex may be divided into hundreds, or even thousands, of small areas, which in the case of functional Magnetic Resonance Imaging (fMRI) investigations are called voxels. The electroencelographic signal is full of artifacts, which is implied by the fact that indicating the source of this signal in human brain may be wrong. The measurement of the EEG signal is taken in compliance with IFCN 10-20 standard (Fig. 4.1), where: A—ear lobe, C—central, P—parietal, F—frontal, T—temporal, O—occipital.

As a result of the EEG investigation, due to using measuring electrodes of adequate resistance, a record of synchronized EPSP/IPSP (Excitatory Postsynaptic Potential/Inhibitory Postsynaptic Potential) neurons is obtained. Carrying out the process of measurement taking based on the electroencelographic investigation, it should be remembered that for various movement visualizations (e.g. of upper and lower limbs, tongue) various areas of the brain become activated. In such cases cortical regions are characteristic of considerable topographic differences. This confirms the validity of taking up the subject matter of the inverse problem in the process of identifying the sources of occurrence of given potentials in human brain, for example the need for optimizing the number of measuring electrodes [4].

© Springer Nature Switzerland AG 2020
S. Paszkiel, *Analysis and Classification of EEG Signals for Brain–Computer Interfaces*, Studies in Computational Intelligence 852,
https://doi.org/10.1007/978-3-030-30581-9_4

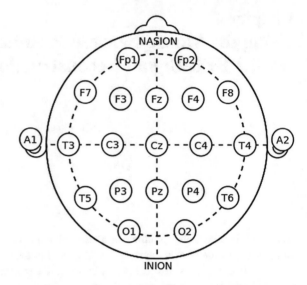

Fig. 4.1 Position of scalp electrodes in compliance with IFCN 10-20 standard

The issue of blind signal separation was widely popularized in the last decade. Here we should provide the names of the following researchers: Jutten and Hérault (1991), Comon (1994) and Sejnowski (1995). Currently, the issue od BSS covers a wide range of applications. It is used for image processing, geophysical data analysis, the analysis of biomedical signals such as: electroencephalographic, electrocardiographic, magnetoencephalographic etc., and also speech analysis. Taking into consideration the use of the electroencephalographic signal analysis, especially in the scope of elimination of artifacts, the following works are of significance: Fitzgibbon et al. (2007), Frishkoff et al. (2007), Halder et al. (2007), Vorobyov and Cichocki (2002).

Analyzing the most simple case, observed signal y(t) is a linear combination of statistically unknown signals independent from one another x(t), which can be expressed with Eq. (4.1), where: a_{ij} is mixing coefficient; $x_j(t)$—input signal, $y_i(t)$—i-measurement signal [1].

$$y_i(t) = \sum_{j=1}^{n} a_{ij} x_j(t) \tag{4.1}$$

In the case of analysis based on BSS there is a possibility of occurrence of single signals in not necessarily the same order as the source signals occurred. Therefore, separation consists in obtaining original source signals but which may occur in a different order than the source signals [7]. Figure 4.2 shows the pattern of mixing signals based on the mathematical Eq. (4.2).

$$y_1(t) = a_{11}x_1(t) + a_{12}x_2(t)$$
$$y_2(t) = a_{22}x_2(t) + a_{21}x_1(t) \tag{4.2}$$

Fig. 4.2 Mixing of input
signals $x_1(t)$ and $x_2(t)$

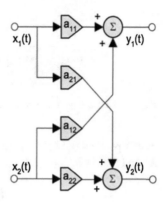

For the purposes of proper identification of signal sources, which is extremely significant from the point of view of measurement taking, a proper mathematical presentation of the above-mentioned sources and the signals read out on the scalp as well as the use of highly specialized tools for the elimination of the signal components that are artifacts is necessary. This was confirmed in the research carried out which proved that the EEG signal may be subjected to the Moore-Penrose pseudoinverse. For example, the system is given by Eq. (4.3), where $\vec{x}$, $\vec{b}$ are vectors and A is matrix. Both A and i $\vec{b}$ are known. Vector $\vec{x}$ is to be found.

$$A\vec{x} = \vec{b} \tag{4.3}$$

Equation (4.3) is transformed from the vector form to algebraic form (4.4).

$$
\begin{aligned}
a_{11}x_1 + a_{12}x_2 + \cdots + a_{1n}x_n &= b_1 \\
a_{21}x_1 + a_{22}x_2 + \cdots + a_{2n}x_n &= b_2 \\
\cdots \\
a_{m1}x_1 + a_{m2}x_2 + \cdots + a_{mn}x_n &= b_n
\end{aligned}
\tag{4.4}
$$

Then, Eq. (4.4) can be used to define matrix form (4.5).

$$
\begin{bmatrix}
a_{11} & a_{12} & \cdots & a_{1n} \\
a_{21} & a_{22} & \cdots & a_{2n} \\
\cdot & \cdot & \cdot & \cdot \\
a_{m1} & a_{m2} & \cdots & a_{mn}
\end{bmatrix}
\begin{bmatrix}
x_1 \\
x_2 \\
\cdot \\
x_n
\end{bmatrix}
=
\begin{bmatrix}
b_1 \\
b_2 \\
\cdot \\
b_n
\end{bmatrix}
\tag{4.5}
$$

Matrix form (4.5) holds true for $m > n$.

$$[A]^{T}[A][x] = [A]^{T}[b] \tag{4.6}$$

After transforming the Eq. (4.6) by moving matrix A and transposed matrix A^{T} to the right side of the equation, X equals (4.7). To that end, it is necessary to ivert the

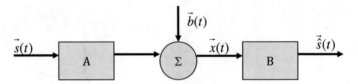

Fig. 4.3 Linear model of mixing EEG signals with estimation

matrix product $(A^\mathrm{T}A)^{-1}$, the resultant matrix is then multiplied by transposed matrix A^T and matrix b.

$$[X] = ([A]^T[A])^{-1}[A]^T[b] \tag{4.7}$$

Notation $(A^\mathrm{T}A)^{-1}\,A^\mathrm{T}$ describes pseudoinverse (left-inverse) written as A^+ (4.8). The pseudoinverse has been developed independently by two scientists, Moore E. H. and Penrose R.

$$[X] = [A]^+[b] \tag{4.8}$$

In order to approximate Eq. (4.1) using pseudoinverse, the relation (4.9) can be determined (4.9).

$$A^+A\vec{x} \approx A^+\vec{b} \tag{4.9}$$

Hence the conclusion (4.10).

$$\vec{x} \approx A^+\vec{b} \tag{4.10}$$

In order to accurately conduct the process of blind signal separation, it is necessary to use for this purpose the Second Order Statistic (e.g.: AMUSE, SOBI, EVD etc.); however, it must be remembered that the components will not correlate with each other. The vector of the signals measured can be defined using the Eq. (4.11).

$$\vec{x}(t) = A\vec{s}(t) + \vec{b}(t) \tag{4.11}$$

where A—mixing matrix; $\vec{x}(t) = [x_1(t), x_2(t), \ldots, x_m(t)]^T$—signal observed on the scalp, $\vec{s}(t) = [s_1(t), s_2(t), \ldots, s_n(t)]^T$—unknown time history of the source, consisting of defined components; $\vec{b}(t) = [b_1(t), b_2(t), \ldots, b_m(t)]^T$—unknown interfering signal—noise. The assumption involves estimation of at least as many sources as there are measuring sensors available at a given moment. Figure 4.3 presents a linear model of the process of mixing electroencelographic signals.

It should be noted that the measurement of the EEG signal, which is written as vector $\vec{x}(t)$ is a linear combination—a mix of source signals written as vector $\vec{s}(t)$. In the case of BBS, we often omit value $\vec{b}(t)$, thus obtaining (4.12).

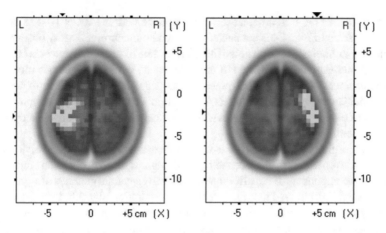

Fig. 4.4 Visualization of source activity using sLORETA method, for the left arm movement on the left side, for the right arm movement on the right side

$$\vec{x}(t) = A\vec{s}(t) \tag{4.12}$$

and after subsequent transformations we obtain an estimated value of vector $\hat{\vec{s}}(t)$ (4.13):

$$\hat{\vec{s}}(t) = B^T \vec{x}(t) \tag{4.13}$$

where: B^T—separation matrix—demixing. Estimated value of vector $\hat{\vec{s}}(t)$ is equal to (4.14).

$$\hat{\vec{s}}(t) = \vartheta \vec{s}(t) \tag{4.14}$$

where

$$\vartheta = B^T A \tag{4.15}$$

There is a correlation between matrix A and transposed matrix B based on the Moore-Penrose pseudoinverse (4.16).

$$A = (B^T)^+ \tag{4.16}$$

Figure 4.4. Shows human brain activity typical for the left arm movement and separately for the right arm movement based on solution of the inverse problem using sLORETA algorithm (standardized Low Resolution Brain Electromagnetic Tomography). For the needs of the experiment carried out, filtration of the EEG signal was conducted in the range from 1 to 45 Hz. Sampling frequency was set to 512 Hz and the number of measuring electrodes was 32. The measurement was taken using Emotiv EPOC+ NeuroHeadset in compliance with the IFCN 10-20 standard. The color scale in Fig. 4.4 is connected with maximum current density observed for given measurements. It can be noted that a high value of the signal power is observed in the left hemisphere during the left arm movement, while the right arm movement causes an increase in the signal power in the right hemisphere. The experiment carried out shows that using sLORETA algorithm we can find only an approximate solution of the inverse problem, with which we deal during electroencephalographic measurements.

BSS algorithms allow for carrying out the process of separation based on information on static properties of electroencelographic signals, and a decision on elimination of selected components of the EEG signal is taken by separate methods of signal analysis. It should be stressed that the selection of separation algorithm should depend on the type and properties of the source signals searched. In order to conduct statistical independence of electroencephalographic signals, one has to assume physical separateness of sources that emit measurable signals. Such an assumption allows to use ICA (Independent Components Analysis) method for separation of the aforementioned signals. This method involves isolation of independent components from multidimensional data set by searching for linear transformation aiming to perform statistical minimization or relation between vector components, while the Moore-Penrose pseudoinverse works perfectly for matrix generalization in the scope of the EEG signal reconstruction after conducting blind signal separation. The experiment carried out using sLORETA method confirms that it is possible to observe differences in brain activity for the particular mental tasks and therefore it makes identification of the generation source of a given potential possible [5].

References

1. Dvorak, I., Holden, A.V.: Mathematical Approaches to Brain Functioning Diagnostics. Manchester University, Manchester (1991)
2. Edla, D.R., Mangalorekar, K., Dhavalikar, G., Dodia, S.: Classification of EEG data for human mental state analysis using random forest classifier. Procedia Comput. Sci. **132**, 1523–1532 (2018). https://doi.org/10.1016/j.procs.2018.05.116
3. Katsikis, V.N., Pappas, D., Petralias, A.: An improved method for the computation of the Moore–Penrose inverse matrix. Appl. Math. Comput. **217**(23), 9828–9834 (2011)
4. PascualMarqui, R.D., Michel, C.M., Lehmann, D.: Low resolution electromagnetic tomography: a new method for localizing electrical activity in the brain. Int. J. Psychophysiol. **18**, 49–65 (1994)
5. Paszkiel, S.: Characteristics of question of blind source separation using Moore-Penrose pseudoinversion for reconstruction of EEG signal, Recent research in automation, robotics and measuring techniques, In: Szewczyk, R., Zieliński, C., Kaliczyńska, M. (eds.) Innovations in Automation, Robotics and Measurement Techniques. Advances in Intelligent Systems and Computing 550, pp. 393–400. Springer, Switzerland (2017). https://doi.org/10.1007/978-3-319-54042-9_36

6. Paszkiel, S.: Laplace filters and blind signal separation for use in the brain computer interfaces. The role of computer science in economic and social sciences. In: Innovations and Interdisciplinary Implications, Kielce, pp. 76–81 (2014)
7. Paszkiel, S.: Moore-Penrose pseudoinversion in terms of identifying the sources of EEG signals. The role of computer science in economic and social sciences. In: Innovations and Interdisciplinary Implications, Kielce, pp. 36–40 (2015). ISSN 2081-478X

Chapter 5
Using the LORETA Method for Localization of the EEG Signal Sources in BCI Technology

Proper localization of a given signal generation can be useful for the construction of precise brain–computer interfaces. In order to obtain that, for the purposes of the conducted works, the LORETA technique was used based on the idea of solving the inverse problems and estimating distribution of electrical activity of neurons in three-dimensional space. This technique is currently frequently used for electrophysiological measurements and its performance has been verified in numerous research laboratories. The following scientists, among others, worked on this issue: Pascual Marqui, Esslen, Kochi, Lehmann in 2002. LORETA is a linear estimation method which does not supplement the EEG signal with new information. Nevertheless, it is owing to this method that estimation of the sources of certain signals in the human brain is possible. They can be then used in the process of control based on Brain–Computer Interfaces (BCI) technology [2].

The issue of properly carried out signal analysis, elimination of artifacts and also correct identification of the sources of their generation, owing to the use of the LORETA technique, is an essential problem in the equipment and software realization of the connection based on BCI.

The inverse problem also called the inverse issue is frequently found in the field of technical sciences. It occurs when some parameters of a certain model have to be determined basing on the values that are possible to observe. In the case of electroencephalography it is a signal measured on the scalp based on the information created in its specific source in a form of the activity of neurons and their mutual correlations. Assuming that $\Omega(t)$ is a set of active dipoles of the signal sources, it is possible to determine the potentials measured on the scalp as (5.1):

$$x(t) = \sum_{i \in \Omega(t)} K_i j_i(t) \tag{5.1}$$

© Springer Nature Switzerland AG 2020

S. Paszkiel, *Analysis and Classification of EEG Signals for Brain–Computer Interfaces*, Studies in Computational Intelligence 852,
https://doi.org/10.1007/978-3-030-30581-9_5

where: i—ith localization of dipoles based on the three-dimensional space: x, y, z, while $x(t) = (x_1(t), ..., x_N(t))^T$ is a vector of data from N of measuring electrodes registered in a moment of time t, K—number of matrices/occurrences of a certain set, j(t)—estimated vector. Dipole is characterized by the changing amplitude and constant orientation.

$$K_q = (k_{q(x)}, k_{q(y)}, k_{q(z)}), \quad K = (K_1, \ldots, K_Q) \tag{5.2}$$

Finally, we are able to express the problem occurring during the measurements with the linear equation (5.3):

$$x(t) = Kj(t) \tag{5.3}$$

where

$$j = (j_1^T, \ldots, j_Q^T)^T, \quad j_q = (j_{q(x)}, j_{q(y)}, j_{q(z)})^T$$
$$q \in \{1, \ldots Q\} \tag{5.4}$$

The fundamental rule during creation of the brain–computer interfaces is proper observation and acquisition of data registered on the scalp. During this operation, it is necessary to localize the sources of potentials in human brain. As it was mentioned in the introduction to this paragraph, in this case we deal with the inverse problem, for the purpose of which, we define the vector j(t):

$$j(t) = T^T x(t) \tag{5.5}$$

where

$$T^T = (T_1, \ldots, T_Q)^T. \tag{5.6}$$

Q—number of voxels in the cortex space, x(t)—potentials measured on the scalp, j(t)—estimated vector for the needs of the inverse problem. Voxel is an element of the volume in the brain space for which we assume a constant value concerning the density and direction of the passing-through current. T^T is an inversion of matrix K.

For the purposes of this experiment, the LORETA method has been character-ized. We can distinguish three kinds of the LORETA techniques for the brain activ-ity imaging—verification of the sources of signals generated in the brain: LORETA, standardized LORETA (sLORETA), exact LORETA (eLORETA). sLORETA is char-acterized by low spacial resolution which decreases with the decrease in the levels of identification. A unique feature of the sLORETA method is a fact of high accuracy of the point sources in ideal conditions. eLORETA is a method which was developed in the University of Zurich, it was extended by the quasilinear methods, owing to which it should be possible to maintain zero number of errors of localization. Nevertheless, as it can be concluded from the research, the sLORETA method is characterized by higher accuracy in the aspect of including biological artifacts than the eLORETA method. The main original matrix of transformation for the sLORETA, that is the standard low resolution electromagnetic tomography, is (5.7):

$$Z = (KK^T + \alpha H)^+ \tag{5.7}$$

where: $Z \in R^{N \times N}$ and is symmetrical. H—centering matrix. A—coefficient owing to which we increase resistance to disturbances. $\Gamma_q(t)$ is defined by the formula presented below (5.8) assuming that $q \in \{1, \ldots, Q\}$.

$$\gamma_q(t) = j_q^T(t)[T^T K]_q^{-1} j_q(t) \tag{5.8}$$

where: $[T^T K]_q$ is the diagonal matrix of the dimensions of 3×3, T^T is a minimal standard of the transposed matrix. Owing to using the sLORETA method it is possible to separate simultaneously active sources of the EEG signal.

The eLORETA method is based on the correlation of the matrix diagonal with its weights. The eLORETA technique is defined by the formula provided below (5.9) in which $q \in \{1, \ldots, Q\}$.

$$\Theta_q^{-1} = [K_q^T (K\Theta^{-1}K^T + \alpha H)^+ K_q^T]^{1/2} \tag{5.9}$$

The issue of the eLORETA method relies on the optimization based on the formula (5.10):

$$\min_{\Theta} \left\| I - [\Theta^{-1}K^T(K\Theta^{-1}K^T + \alpha H)^+ K\Theta^{-1}] \right\|_F^2 \tag{5.10}$$

which is met by the following formula (5.11), where $q \in \{1, \ldots, Q\}$

$$\Theta_Q^2 = K_q^T(K\Theta^{-1}K^T + \alpha H)^+ K_q \tag{5.11}$$

Θ is determined by the matrix (5.12).

$$\Theta^{-1} = \begin{bmatrix} \Theta_1^{-1} & & 0 \\ & \ddots & \\ 0 & & \Theta_Q^{-1} \end{bmatrix} \tag{5.12}$$

Data for the experiment were collected in the NeuroScience Laboratory of the Opole University of Technology. Five persons (22 years old on average) on whom the measurements were taken were in a sitting position. The measurement lasted 5 min, it was repeated 30 times for each state: elevated concentration—solving mathematical problems, organism's calmness—closed eyes, processing standard information from the surroundings—eyes open. The signals were measured in compliance with the IFCN 10–20 standard specified by the International Federation of Clinical Neurophysiology. Emotiv EPOC+ NeuroHeadset was used as a measuring device. After taking signal samples, they were filtered using the software with which biological artifacts such as nervous tics of the examined persons, temporary clenching eyelids, etc. were eliminated. Within the experiment carried out, one matrix for each waveband: alpha, beta, delta, theta was determined. A rectangular window for analyses was specified based on the research by Lubar, Congedo, Askew conducted in 2003. During the research conducted, the LORETA-Key was used. The measurement space was determined based on the atlas of the Brain Imaging Center at the Montreal Neurological Institute, through dividing the brain into 2394 voxels of the dimensions of: $7 \times 7 \times 7$ mm.

For the purposes of the conducted data analysis, each voxel of the fixed current density was standardized in the range of amplitude up to 9 mm using a three-dimensional filter, and then subjected to logarithmic transformation in order to approximate the data using the Gaussian function. In the case of using the LORETA technique, it is important to decrease the errors caused by the anatomical structure of the head and errors caused by the incorrect localization of the sources. Unfortunately, local maxima of the function can be shown in other places than they really occur. Then, the spatial normalization bases on the process of normalization of the square root from the sum of the squares of the current density. Owing to using the above-mentioned mathematical apparatus it is possible to eliminate incorrect values introduced by, for instance: different skull thickness for particular persons, differences in the impedance of electrodes, etc. Estimation of the current amplitude density makes it possible to provide data for further statistical analysis. The experiment was carried out for the waves: alpha, beta, theta and delta, voxel after voxel, using the t-test. For the above-mentioned wavebands, the LORETA technique was used for the analysis of the current amplitude density.

Figures 5.1 and 5.2 show exemplary visualizations of the changes in activity of certain areas of the brain activity of human brain using standardized Low Resolution Electromagnetic Tomography. Figure 5.1 shows the activity of the sources of electroencephalographic signals for elevated concentration, which in the case of implementation in the scope of BCI technology, may be a signal implying a movement of a robot. Figure 5.2 shows the activity of the EEG signal sources for the state of processing standard information from the surroundings of the examined persons,

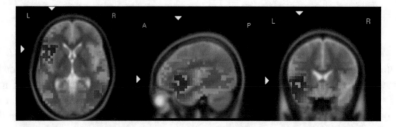

Fig. 5.1 Exemplary visualization of the brain activity based on sLORETA for the state of elevated concentration

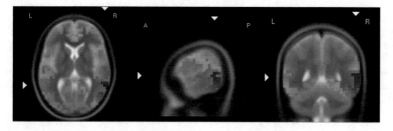

Fig. 5.2 Exemplary visualization of the brain activity based on sLORETA for the state of processing standard information from the surroundings

which in the case of implementation may be the information implying stopping a robot.

For the purposes of the research, sLORETA method was chosen for imaging brain activity in relation to three states: concentration, calmness, standard processing. Properly conducted identification of the electroencephalographic signal sources using the sLORETA technique made it possible to obtain information on creation of specific changes in the EEG signal readings for specific measuring electrodes. Changes in the amplitude of signals correlated with specific sources of their generation in human brain observed in the above-mentioned group of people in relation to the tested states of concentration can be successfully used to execute control with brain-computer technology. LORETA enables the execution of visualization of incorrect activity of brain structures determining, at the same time, artifacts within the range of frequency band and special localization, which enables the performance of much more advanced analyses than the standard data analysis based on electroencephalography [1]. Owing to the application of the LORETA technique, it is not necessary to limit the number of bipolar point sources, because the current distribution on the entire brain area is calculated directly. Then we obtain a three-dimensional tomographic image, received assuming a low value of spatial resolution. It should be noted that owing to using the LORETA method it is possible to specify new hypotheses concerning cognitive functions performed by the brain. The Common Spatial Pattern is an alternative method used for spatial localization of signal sources, which is also worth considering as far as control process support based on BCI technology is concerned [3].

References

1. Lagerlund, T.D., Sharbrough, F.W., Busacker, N.E.: Spatial filtering of multichannel electroencephalographic recordings through principal component analysis by singular value decomposition. J. Clin. Neurophysiol. **14**, 73–82 (1997)
2. Paszkiel, S., Hunek, W., Shylenko, A.: Project and simulation of a portable proprietary device for measuring bioelectrical signals from the brain for verification states of consciousness with visualization on LEDs, Recent research in automation, robotics and measuring techniques. In:

Szewczyk, R., Zieliński, C., Kaliczyńska, M. (eds.) Challenges in Automation, Robotics and Measurement Techniques. Advances in Intelligent Systems and Computing, vol. 440, Conference Proceedings Citation Index (ISI Proceedings), pp. 25–36. Springer, Switzerland (2016). https://doi.org/10.1007/978-3-319-29357-8
3. Paszkiel, S.: Using LORETA method based on the EEG signal for localising the sources of brain waves activity for the purpose of brain-computer interfaces. Meas. Autom. Monit. (MAM) **62**, 262–264 (2016). (ISSN 2450-2855)

Chapter 6
Data Analysis of Human Brain Activity Using MATLAB Environment with EEGLAB

For the purpose of the research work carried out, Emotiv Xavier TestBench environment was used and the biomedical data obtained from it may be used for further processing. The data format (*.edf) is compatible with Toolbox EEGLAB [1]. It is an interactive tool within the Matlab environment for processing continuous data connected with EEG, MEG events and other electrophysiological data covering Independent Components Analysis (ICA), time analysis, frequency and artifacts removal. EEGLAB operates under Linux, Unix, Windows and Mac OS X systems. Toolbox EEGLAB is not available in the default version of Matlab, therefore it must be installed. Figure 6.1 presents an EEGLAB configuration window, where data processing is conducted. Figure 6.1 also delineates an access path for reading the obtained data.

It can be observed that a structure occurred within the working space where all the parameters are incorporated. Raw data themselves are stored in the matrix form (Fig. 6.2), from which the data from subsequent electrodes are provided in the following verses and the data from subsequent time spans—in columns.

It can be observed that the data in Fig. 6.3 are complex for identification and analysis. There are a lot of disturbances (artifacts). Thus, the EEGLAB environment enables to process the data in order to gain e.g. only the specified signal frequencies. According to theory knowledge on identification of oscillations in human brain [199], it is well-known that the beta waves frequency ranges from 13 to 30 Hz, therefore a band-pass filter has been used in order to obtain only this signal on the individual electrodes. The configuration of this filter is presented in Fig. 6.4.

The use of the band-pass filter causes another data object to occur which was generated as a result of passing a primary signal through the above-mentioned filter (Fig. 6.3).

Figure 6.5 also presents Bode's plot characteristics for the applied filter. It shows that for the range 13–30 Hz, both the module and the signal phase change.

Figure 6.6 shows the obtained signal transformed by the filter. As a result of comparison of Figs. 6.3 and 6.6, it is possible to realize that a band-pass filter is an

© Springer Nature Switzerland AG 2020

S. Paszkiel, *Analysis and Classification of EEG Signals for Brain–Computer Interfaces*, Studies in Computational Intelligence 852, https://doi.org/10.1007/978-3-030-30581-9_6

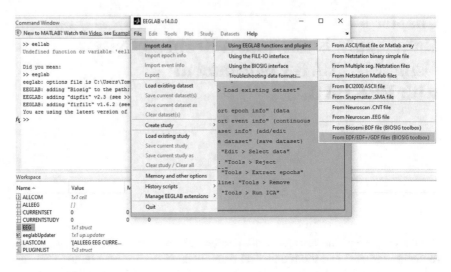

Fig. 6.1   EEGLAB window for Matlab

excellent device for artifact filtration, disturbances having impact on the obtained
signal. The data prepared in this way may by used for controlling various processes,
which will be presented in further chapters of this monograph. Figure 6.7 shows
further exemplary data. It can be observed that a new structure occurred in the working
space, which contains all the parameters. Raw data are stored in the matrix form, for
which the data from subsequent electrodes are provided in the following verses and
the data from subsequent time spans—in columns.

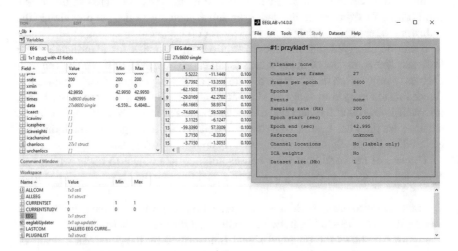

Fig. 6.2   Input data—Matlab

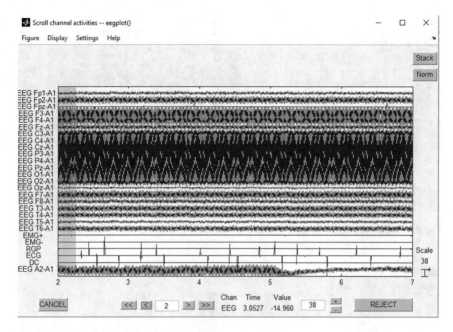

Fig. 6.3 Received data characteristics (without processing)

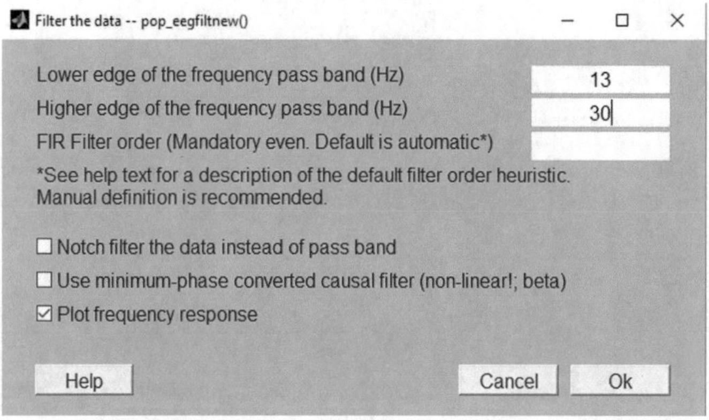

Fig. 6.4 Band-pass filter configuration for f: 13–30 Hz

As it can be observed in Fig. 6.8, the data are distorted to a large extent. There are a lot of disturbances (artifacts). However, the Matlab environment makes it possible to process these data to obtain, for example, only definite signal frequencies. It is known that the beta waves frequency ranges from 13 to 30 Hz, therefore a band-pass filter has been used in order to obtain only this signal on individual electrodes.

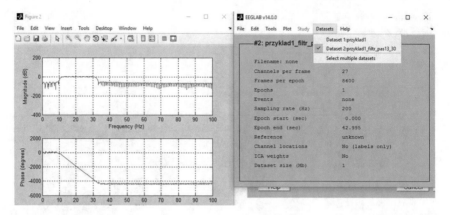

Fig. 6.5 Band-pass filter configuration for f: 13–30 Hz

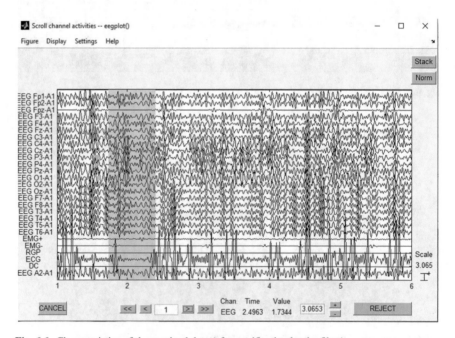

Fig. 6.6 Characteristics of the received data (after verification by the filter)

The use of the band-pass filter causes the occurrence of another data object, which was generated as a result of passing a primary signal (Fig. 6.8) through the above-mentioned filter (Fig. 6.9).

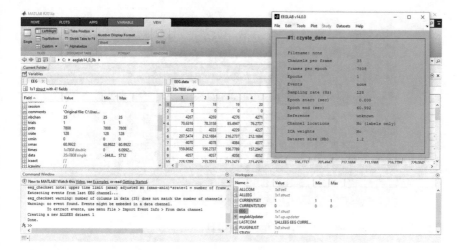

Fig. 6.7 Input data in Matlab environment—example 2

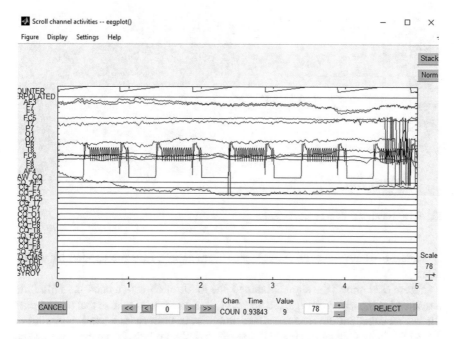

Fig. 6.8 Characteristics of the received data (without processing)—example 2

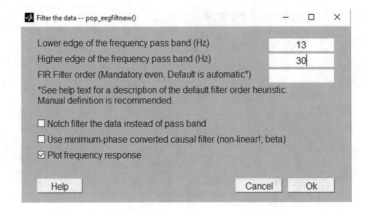

Fig. 6.9 Band-pass filter configuration for f: 13–30 Hz

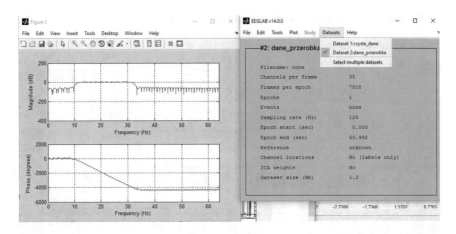

Fig. 6.10 Band-pass filter for f: 13–30 Hz

Figure 6.10 also presents Bode's plot characteristics for the applied band-pass filter.

Figure 6.11 shows the signal obtained due to filtration with a band-pass filter. As a result of comparison of Figs. 6.8 and 6.11, it is possible to realize that a band-pass filter is an excellent device for artifact filtration. In both cases a band-pass filter of 13–30 Hz frequency values made it possible to obtain relatively clear signals which might be used for further processing or for application in a control process. The data format (*.edf) is a commonly used format for data archivization, which allows a fast development of devices of various manufacturers and their relatively easier mutual

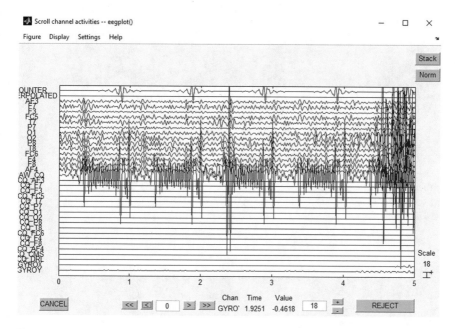

Fig. 6.11 Characteristics of the received data (after verification by the filter)

compatibility. Owing to taking advantage of theory, including, among others, the knowledge of the frequencies of the particular waves and adequate filters eliminating artifacts, a more and more accurate reading of the human brain activity states becomes possible, which translates into wider and wider application possibilities [2].

References

1. Delorme, A., Makeig, S.: EEGLAB: an open source toolbox for analysis of single-trial EEG dynamics including independent component analysis. J. Neurosci. Methods **134**(1), 921 (2004). https://doi.org/10.1016/j.jneumeth.2003.10.009
2. Paszkiel, S., Szpulak, P.: Methods of acquisition, archiving and biomedical data analysis of brain functioning, Biomedical Engineering and Neuroscience. In: Hunek, W., Paszkiel, S. (eds.) Proceedings of the 3rd International Scientific Conference on Brain-Computer Interfaces (BCI 2018), Opole, Poland, 13–14 Mar 2018. Advances in Intelligent Systems and Computing (AISC), vol. 720, pp. 158–171. Springer, Berlin (2018). https://doi.org/10.1007/978-3-319-75025-5_15

Chapter 7
Using Neural Networks for Classification of the Changes in the EEG Signal Based on Facial Expressions

Artificial intelligence dates back to 1950, when a group of pioneers in informatics began asking questions related to the ability "to think" by computers defined as automation of intellectual tasks performed by human beings. The consequences of the questions asked by the pioneers of informatics have been discovered until now in the form of many scientific and technological achievements. Artificial intelligence in itself is a very general discipline, which includes the issues from such fields of knowledge as machine learning and deep learning, and also numerous other approaches that do not require learning.

Symbolic artificial intelligence is one of the types of artificial intelligence which turned out to be suitable for solving precisely defined and logical problems; however, it was not applicable to more complex tasks such as speech recognition, language translation or image classification. With the development of works on artificial intelligence, a new approach to the issue of artificial intelligence called machine learning developed. With technological development and progress in implementation of IT solutions, creating a computer which will go beyond processing responses based on data and the rules specified by a user. The questions were asked whether a computer, like man, can learn and independently solve complex logical problems. Answers to these questions opened the door for a new programming paradigm. In classical software development, programmers enter data with a set of rules, and a definite response is received on output. In a new concept of programming presented in Fig. 7.1, based on data and responses, the computer was able to define the rules taking place between the entered data. In this way a new branch of artificial intelligence, called machine learning, came into existence.

Machine learning is a field of artificial intelligence which uses algorithms for the synthesis of basic dependencies between data and feedback information. For example, machine learning systems can be trained on speech recognition systems transforming sound information to a speech data sequence into a semantic structure expressed in the form of a sequence of words. Presently, machine learning systems are widely used in software for speech recognition, smart houses, controlling robots,

© Springer Nature Switzerland AG 2020
S. Paszkiel, *Analysis and Classification of EEG Signals for Brain–Computer Interfaces*, Studies in Computational Intelligence 852,
https://doi.org/10.1007/978-3-030-30581-9_7

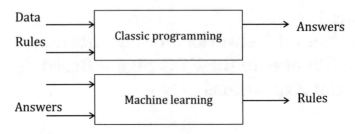

Fig. 7.1 Machine learning as a new programming paradigm

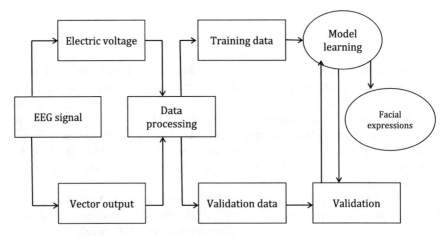

Fig. 7.2 Diagram of supervised learning for facial expression identification

autonomous vehicles and in the analysis of large data sets and other intelligent applications. Machine learning owes its growing popularity to its ability to characterize
complex relationships and thus, with the development of *big data* it can determine
correct structures and behavior patterns for the issues significant for contemporary
man.

Machine learning is a vast field of knowledge, where for a given task a kind of
a problem must be defined and proper algorithms must be used for their solution.
These algorithms are divided into four categories: supervised learning, unsupervised
learning, semi- supervised learning, reinforced learning. The mechanism of supervised learning shown in Fig. 7.2 defines correlation between the observed input data
and the target variable being subject to prediction. The task of learning uses labelled
data for the synthesis of a model which attempts to generalize relationships between
feature vectors and supervisory signals. The feature vector influences the direction
and size of a change in such a way as to improve general performance of the learning
model.

A well-trained predictor model based on the supervised algorithm can precisely
predict output data for unseen phenomena taking place in unknown and unobservable

data instances. The aim of learning algorithms is minimization of an error for a given training set. Learned data of a low quality have influence on observation accuracy and comprehensiveness, in consequence leading to overfitting which is a poor generalization and data misclassification. Unsupervised learning possesses mechanisms for detecting unseen structures in unlabeled data sets without output data. Unsupervised learning has found many applications in the scope of detecting outstanding dependencies of data compression. Both supervised unsupervised and supervised learning are not strictly defined terms due to their unclear division. This division results from the fact that that various learning techniques may be used both for the former and the latter approach. For example, the chain rule defines decompositions of common distributions for a random input vector x E R^n and is given by Eq. (7.1):

$$(x) = IIp(x_i | x_1, \ldots, x_{i-1}) \tag{7.1}$$

It means that there is a possibility of solving the problem of unsupervised modelling $p(x)$ by dividing it into n supervised problems. Additionally, the problem of supervised learning $p(y|x)$ may be solved using unsupervised learning methods to get to know the common distribution $p(x, y)$ and then the inference:

$$p(x|y) = \frac{p(x, y)}{\sum_{y'} p(x, y')} \tag{7.2}$$

Both algorithms are neither fully formalized nor totally separate but they help in solving the tasks performed using artificial intelligence algorithms. Semi-supervised learning uses the combination of a small number of labelled and a big number of unlabelled data to generate the model function. Semi-supervised learning has been categorized between the guidelines for supervised learning and the guidelines for unsupervised learning. This method became very popular in learning electronic, retail, biological etc. data. It is also of practical value in the fields connected with man, such as speech recognition or handwriting representation. Another type of learning is reinforcement learning, which consists in such representation of input data as to maximize the signal of reward. The learning algorithm does not have information on the tasks to be performed but it must find out which actions bring the biggest reward using the trial and error method. In most interesting and most demanding cases, the process of observation may have influence not only on the increases of the reward coefficient but also on the quality of subsequent observations.

7.1   Machine Learning Problems

The main aim of classification is categorization of data based inputs x and outputs y where e $y \in \{1, \ldots, C\}$. Depending on the number of C classes, classification task is divided into: binary classification where $C = 2$ i $y \in \{0, 1\}$, multiclass classification where $C > 2$ i $y \in \{0, 1\}$, multilabel classification for C denoting the number of classes. If the number of classes is two, the problem is the so-called binary classification where $y \in \{0, 1\}$). If the number of classes is $C > 2$, the problem is the so-called multiclass classification. In the case in which the labels of classes are not mutually exclusive, the problem refers to multilabel classification. One of the ways of solving classification problems is approximation of its function. Assuming that $y = f(x)$ for indefinite function f, and the aim of learning is estimation of function f using trained data and then to perform prediction using $y = f(x)$. The main aim of classification is predicting new input data which have not been observed before.

The task of regression is to estimate the output value based on the input vector. As a result of a proper solution of the regression task, the learning algorithm should obtain the function $f: R^n \rightarrow R$. The regression task is similar to classification task but the difference is that for classification the output variable is categorical and in regression numerical. In machine learning, due to the type of the output value, we distinguish: scalar regression where the aim is to obtain a continuous scalar value, and a vector regression where we obtain a vector of continuous values. Due to the types of regression we distinguish linear regression and logistic regression. The aim of linear regression is to create a system assuming input vector $x \in R^n$ and predicting output scalar value $y \in R$. As a result we obtain a linear function the independent variable of which are input data. The result is defined as:

$$y^\wedge = w^T x \qquad (7.3)$$

where: $y^\wedge$—value predicted by the model; $w \in R^n$—set of weights. The parameter vector contains values responsible for the behavior of the system performed. Parameter w_i is a coefficient which is multiplied by x_i feature still before all features have been summed up. A set of weights w defines what influence a given parameter has on future prediction. Assuming that: $w_i > 0$, causes increase x_i is directly proportional to predictions $y^\wedge$, $w_i < 0$, causes increase x_i is inversely proportional to predictions $y^\wedge$. If a big weight coefficient is assigned to a given feature, this feature has a significant influence on the final prognosis. Otherwise (when the weight coefficient of a feature is close or equal to 0), it has no influence on prognosis. In order to improve the performance measure of the model made, the objective function, which is responsible for error minimization, is used. In the case of linear regression, the mean squared error for test data has been calculated expressed in Eq. (7.4).

$$MSE = \frac{1}{m} \sum (y^\wedge - y)^2 \qquad (7.4)$$

The equation shows that the error measure decreases to zero when $y^\wedge = y$, and also that the error value is directly proportional to the difference between the predicted value and the output value.

Logistic regression defines a probabilistic model which predicts probability of a given event occurrence. The logistic regression model specifies correlation between the vector of input values **x,** and the predicted value y ∈ {0, 1} which is expressed in the form of probability. The logistic regression function can be expressed with Eq. (7.5).

$$P(X|Y) = \frac{e^{B_0 + B_1 x}}{1 + e^{B_0 + B_1 x}} \tag{7.5}$$

The logistic regression function can also be transformed to the inverse of the logistic function called the logit function which is a key to generating coefficients of the logistic regression function. This transformation is represented with Eq. (7.6).

$$\log it(P(Y|X)) = \ln \frac{P(Y|X)}{1 - P(Y|X)} \tag{7.6}$$

The logistic regression function provides the best fit to the curve using regression coefficients $B_0 + B_1$ expressed with Eq. 7.5 where the output response is given in dichotomous scale and the input value is a numeric variable. Due to the fact that the logistic function curve is nonlinear, the logit transformation expressed with Eq. 7.5 is used linearly, where probability distribution (l) is expressed by the degree of probability of the output value for a given input value.

7.2 Deep Learning

Deep learning as an area of machine learning is a new approach to learning data representations. In deep networks stress is put on training subsequent layers increasing the significance of representations. These networks base on the idea of successive adding of the layers, however scientific literature does not determine how many layers are required to claim that a given networks is a deep network.

State-of-the-art deep learning methods include tens and hundreds of successive layers which learn automatically until data exposition and division into training data. For comparison, other approaches to machine learning focus on one or two layers and therefore they are commonly referred to as "shallow". Figure 7.3 shows an example of a deep neural network model.

In deep neural networks, data representations are almost always trained using the structures called *neural networks*. This term refers to neurology, however, in spite of the fact that some of the main concepts of deep learning drew inspiration from human brain, they are not directly its models. The characteristic feature of the layers in neural networks is storing weights in the form of a set of numbers. From the point of view of informatics, the transformation implemented by a layer consists in parametrization of weights of the particular neurons. In this case learning means finding a proper set of weights and values in such a way that the network may

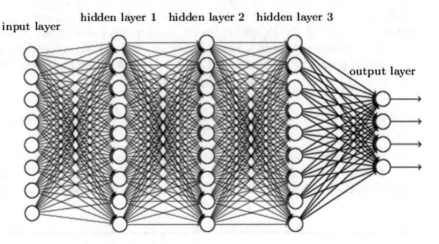

Fig. 7.3 Neural network model containing a few deep layers

properly map training data on the input to responses or goals connected with them. The characteristic feature of deep neural networks is the fact that the network based on the deep structure may contain tens of millions of parameters. During initiation, weights have assigned random values close to zero so the network implements only a series of random transformations.

Cross-validation is the method often used for determining proper parameters. It is performed by splitting learning data D into equal K folds marked as $D_j, j = 1, …,$ L. In each step of this method D_j is used as a test dataset and the other $K - 1$ folds are used as training data. Basing on predicted results and test values, quality index r_j is determined. The cross-validation method is shown in Fig. 7.4.

Having performed all the steps of the cross-validation test, the average value of prediction r can be determined for the whole set. The aim of the cross-validation test is finding a set of parameters that will maximize the average value of prediction index r. The k-fold cross validation algorithm is used when the complexity of D data set is low for test and training data or when the average of losses may have a too big

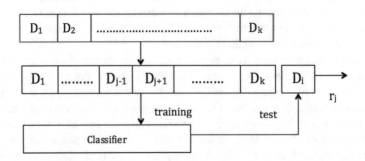

Fig. 7.4 Cross-validation method

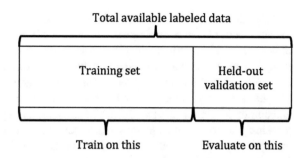

Fig. 7.5 Breakdown of data using the hold out method

variance. Another validation method is the validation of *hold out* type. Assuming that all data points are independent and of identical distribution, a dataset of training data is determined randomly. The example of this breakdown is shown in Fig. 7.5.

In the case of this method validation data usually do no exceed 30% of training data. The model is trained on a bigger part of data and then a validation measure coming from prediction on a smaller data subset is determined. Computation-wise this method is easy to implement and quick to execute. Validation results in the *hold-out* method come from a decreased dataset therefore it should be taken into account that the generalization error may be less reliable. It is best to use the *hold-out* method when the number of data is big enough to ensure reliability of the generalization error. One of the most common challenges of the machine learning algorithms is the possibility to receive the best quality index on testing data. This index when obtained on learning solely on training data is not always reliable in the context of newly received data. Machine learning algorithms can, without any major problems, predict dependencies taking place for a given dataset, but the situation in which an algorithm extracts the features and dependencies characteristic only and solely for training data is an unwanted phenomenon. The aim of the learning algorithm is to analyze the training data in such a way as to be able to make generalizations with which it will successfully manage new data. In other words, it is recommended that the learning error on testing data, called the test error, be as low as during the training error.

In order to influence the performance of learning on the test set we should refer to the theory of statistical learning, which says that the data subjected to the process of learning must meet the following requirements: The training test and the test set cannot be created in an arbitrary way; The examples in datasets must be independent from one another; The test set and the training set must have an identical distribution received from the same probability distribution. In the next step it is necessary to define the factors determining when the algorithm will act best for a given problem. One of the factors is the ability to receive the training error of the smallest error possible. Another factor is obtaining the smallest possible difference of the error between training and test errors. The phenomenon in which an algorithm cannot obtain a sufficiently small value of error for a given dataset is called misfitting. The phenomenon in which the difference between the training and test error is too big is called overfitting.

In order to minimize overfitting the following solutions can be used: increase of the number of data in the training set, reduction of the model capacity, regularization of weights, random dropout. Model capacity is defined as the ability to fit in relation to various functions. The problem of misfitting is often correlated with a low capacity of a model. Excessive capacity may cause a big difference between the learning error and the test error. Capacity may be manipulated by changing the number of input and output data in this model. One of the main principles of economy and economic thinking, the so-called *Ockham's razor*, says that from many hypotheses describing correctly a given phenomenon, the hypothesis which has the least assumptions should be chosen. This also refers to learning models where complex network models are more susceptible to overfitting than less complex models whose distribution of parameters has a smaller number of states. In order to mitigate the model overfitting, constraints on the network complexity are imposed through limiting the weight values to minimum, which results in a more regular weight distribution. In practice, this solution is implemented as added the cost function connected with having big weights. The additional cost is taken into account in two ways: Regularization $L1$—added cost is proportional to the absolute value of weight coefficients; Regularization $L2$—the added cost is proportional to square value of weight coefficients.

Learning using neural networks with many parameters is a very efficient method of machine learning. However, overfitting is a very serious problem in this type of networks. The downside of neural network is also a slow learning, which makes it difficult to deal with overfitting by combining predictions in many various neural networks during a test. The method of random dropout is one of the ways of solving the problem of overfitting. The key feature is a random dropout of neurons together with their connections during the process of learning, owing to which the risk of overfitting decreases. An example of operation of the random dropout is presented in Fig. 7.6.

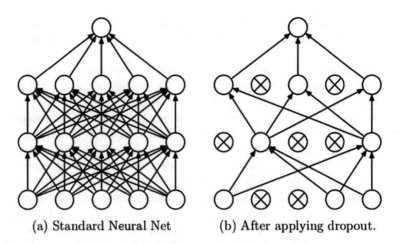

(a) Standard Neural Net   (b) After applying dropout.

Fig. 7.6 Example of random dropout application

During learning a neural network removes neurons from the exponential number of various networks. During a test it is easy to estimate the influence of averaging prognoses of all these thin networks simply using a single network without connections, which has smaller weights. This decreases overfitting significantly and provides a considerable improvement in relation to other methods of regularization.

7.3 Neural Network Implementation

Keras library was the main tool during the neural network implementation for the needs of the research carried out. It is a library written in *Python* for creating the models of artificial neural networks. Additionally, for the purpose of data preparation for training a neural network, such libraries as: *ScikitLearn, Pandas and Matplotlib* were used for data processing and generating diagrams.

The data were collected using a noninvasive method of the EEG signal measurement using Emotiv EPOC+ Neuroheadset operating within the scope of BCI technology. The scheme of localization of electrodes during the test complied with 10-20 standard. The data from the process of data acquisition were registered using *Emotiv TestBench* software. During data collection, a person performed such actions as: eye blinking, teeth clenching, smiling or raising eyebrows. These actions caused the appearance of artifacts of biological nature in the EEG signal based on which the classification of EEG signals [1] using a neural network was carried out. The process of creating a neural network was carried out in the following way: (1) Operations on data: getting acquainted with the data; separation of categories; generation of output data matrix; normalization. (2) Neural network design: defining a structure, selection of parameters (3) Training: performance of the process of training; validation.

At the first stage the person carrying out the research got acquainted with the data coming from biological artifacts. The data contained the values of electric voltage expressed in uV for the electrodes localized on the head. The names of electrodes and their location are shown in Table 7.1.

Sample values for the data from the EEG signal are shown in Tables 7.2 and 7.3.

Based on the data from the measuring electrodes, the EEG signals with the following artifacts were generated: teeth clenching Fig. 7.7, smile Fig. 7.8, closing eyes Fig. 7.9, raised eyebrows Fig. 7.10, no artifacts Fig. 7.11.

As it is shown in the figures above, significant changes in the signal amplitude denoting the performance of an action by the person controlling the device took place due to generation of biological artifacts. It can also be read from the character-

Table 7.1 Designation of the EEG signal input data

Left hemisphere	AF3	F7	F3	FC5	T7	P7	O1
Right hemisphere	AF4	F8	F4	FC6	T8	P8	O2

Table 7.2 Sample voltage values on the electrodes

	AF3	F7	F3	FC5	T7	P7	O1
0	4151.261738	4166.666504	4173.333008	4163.076660	4179.487305	4171.794922	4171.794922
1	4138.974121	4147.692383	4150.256348	4142.051270	4166.153809	4162.051270	4146.153809
2	4132.307617	4138.974121	4140.512695	4127.692383	4155.897461	4156.410156	4144.615234
3	4128.205078	4142.051270	4136.410156	4132.820312	4155.897461	4156.922852	4143.589844
4	4136.922852	4152.307617	4143.076660	4150.256348	4168.205078	4174.871582	4152.820312

Table 7.3 Sample voltage values on the electrodes

O2	P6	T8	FC6	F4	F8	AF4
4173.846191	4173.846191	4170.256348	4170.256348	4155.384277	4163.076660	4162.563965
4149.743652	4144.615234	4139.487305	4153.846191	4133.333008	4145.127930	4142.563965
4144.615234	4130.769043	4125.641113	4145.127930	4128.205078	4133.333008	4138.461426
4152.307617	4138.461426	4126.666504	4148.717773	4130.256348	4138.974121	4142.051270
4165.641113	4154.871582	4134.871582	4163.076660	4136.410156	4152.820312	4147.692383

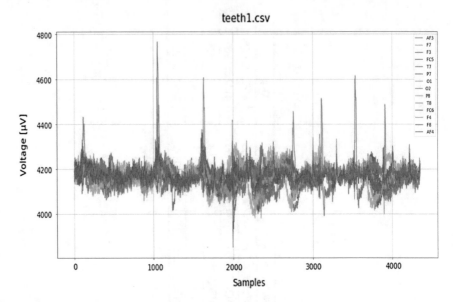

Fig. 7.7 EEG waveform for the artifact connected with teeth clenching

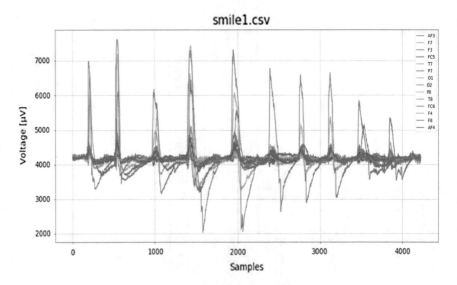

Fig. 7.8 EEG waveform for the artifact connected with a smile

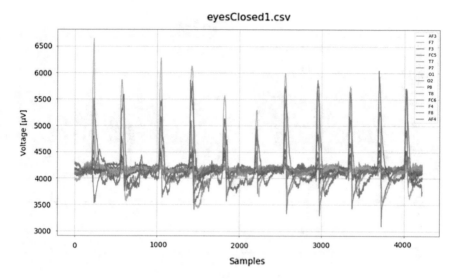

Fig. 7.9 EEG waveform for the artifact connected with closing eyes

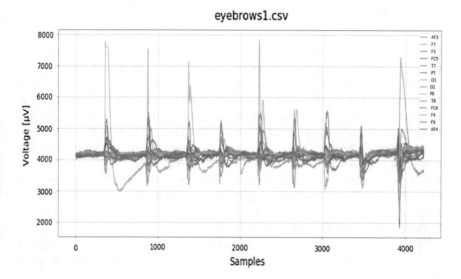

Fig. 7.10 EEG waveform for the artifact connected with raising eyebrows

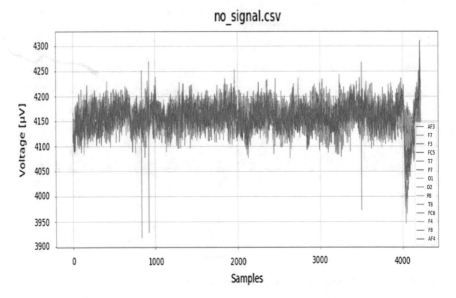

Fig. 7.11 EEG waveform in the case of lack of an artifact

istics that with increasing frequency of the signal occurrence, the value of inferences decreases, nevertheless, they are observable in comparison with a standard waveform.

The values of electrodes change in time, but due to the selected type of the neural network and the type of machine learning, time will not be considered. Due to the absence of the output vector in the form of a class number, it might seem that the neural network designer has to solve the problem of unsupervised learning. Based on the chain rule, the problem of unsupervised learning was solved through splitting it into n supervised problems where each problem denoted a different class. For this purpose a measurement was taken when the person tested did not perform any action. Based on this measurement, an index was generated according to which an artifact occurrence was checked. Figure 7.12 shows a program code for determining the threshold value of the class "no artifact".

Based on the index a program was executed with which an output vector for each of the classes was created. Figure 7.13 shows a program code for generating output vectors.

As it can be observed on the figure above, the number of data for training contains over 57,000 records divided into 15 columns, 14 of which are the electrode values and the last one is the vector of output values. The quality index was determined by recording the highest amplitude value of the signal coming from the generated waveform. Because, for the purpose of the research work, the signal of signal occurrence was determined independently, all records connected with lack of a signal were removed (Fig. 7.14).

```
10   ### Generating highest value for class 'lack of artifacts'
11
12   no_signal = pd.read_csv("no_action.csv")
13   no_signal = no_signal_iloc[: ,2:16]
14   no_signal_max = 0
15
16   for k in no_signal[:].values :
17       for l in k:
18           if l > no_signal_max :
19               no_signal_max = 1
20   print("Max class value 'lack of artifacts' : ", no_signal_max)
```
Maximum value for class '0' : 4332.307617187501

Fig. 7.12 The way of calculating maximum value for the class "no artifact"

```
20   eegCSV = ["eyesClosed2.csv","eyebrows2.csv","smile2.c:
21   dataArray = []
22   counter = 0
23   for j in range(len(eegCSV)): #csv
24       current = pd.read_csv(eegCSV[j])
25       current = current.iloc[: ,2:16]
26       current["Output"] = 0 #ADD output column
27       for x in current[:].values : # row
28           for l in x: #numbers
29               current.iloc[counter,-1] = j+1
30           counter = counter +1
31           #print(counter)
32       dataArray.append(current)
33       counter = 0
34
35   eegToLearn = pd.concat(dataArray,ignore_index=True)
36   eegToLearn = eeg.ToLearn[eegToLearn.Output != 0]
```
Amount of data to learn (57472,15)

Fig. 7.13 Program generating an output vector

As it can be observed, the number of data for training decreased dramatically. This is caused by a high sampling frequency and the fact that the brain reacts only to the fact of a stimulus occurrence, which lasts for a very short period of time. Table 7.4 shows classes with the numbers assigned to them.

```
25 eegToLearn = eegToLearn.Output != 0]
26 print("Amount of data to learn" :",eggToLearn.shape)
```

Amount of data to learn: (7607,15)

Fig. 7.14 Number of data after eliminating the class "no signal"

Table 7.4 Assignment of classes to the corresponding numbers

The name of the class	The assigned number
Eyes closed	1
Raised eyebrows	2
Smile	3
Teeth clenching	4

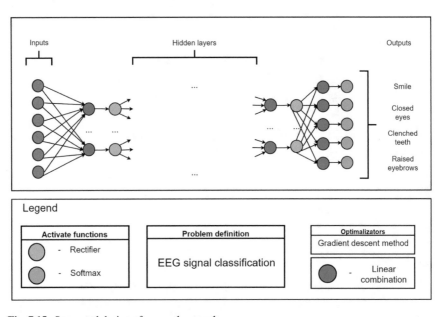

Fig. 7.15 Suggested design of a neural network

At the very end of the stage of getting acquainted and data processing, data vectorization and division into the training set and test set were performed. The test set constitutes 20% of all possible data. Figure 7.15 shows a suggested design of creating o neural network. Below, there is a legend defining applied activation functions in a given neuron layer and the optimizers used.

In the next stage, a graph of neural networks with the description of the particular layers was presented. This graph is shown in Fig. 7.16.

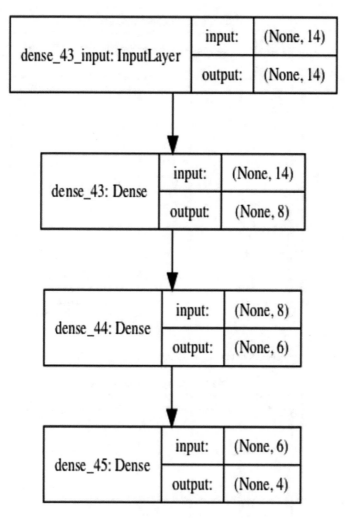

Fig. 7.16 Graph showing the final structure of the neural network

As it is shown in the graph above, the neural network consists of one input layer, two unseen layers and one output layer. The values of 14 electrodes were attached in the input layer, which during training were going through the unseen layers. The unseen layers were activated using the *RelU* method and the output layer using the *Softmax* function. During training, the network was validated using the "Hold Out" method. The characteristics of the neural network training are shown in Figs. 7.17 and 7.18.

As it is shown in Fig. 7.17, the value of precision during the process of learning with validation achieved a very high and satisfying level of 98%. Furthermore, it can be stated that the dependencies between the electrodes are very easy to be determined by the algorithms of artificial intelligence due to the fact that the model achieves a very

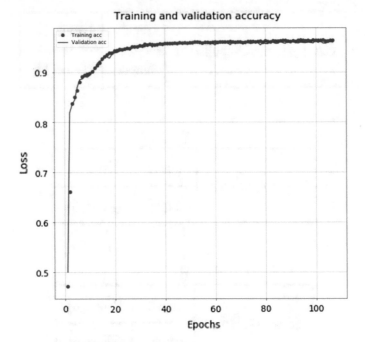

Fig. 7.17 Accuracy coefficient of the model during training and validation

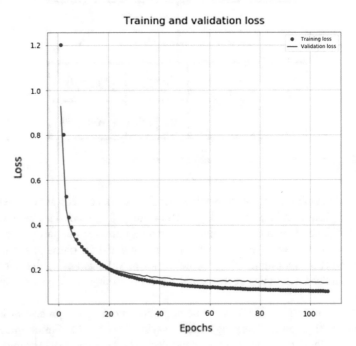

Fig. 7.18 Loss coefficient of the model during training and validation

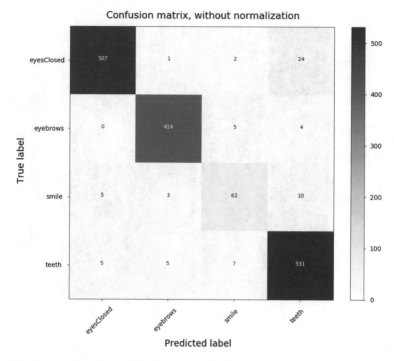

Fig. 7.19 Error matrix of the model without normalization

high level of prediction already after 20 *Epochs*. It can be read from the characteristics of the loss function of the model that overfitting did not occur during training, which confirms a good generalization of the model created. In order to verify the correctness of prediction on the data never seen by the network in the process of learning, error matrices were presented. They are shown in Figs. 7.19 and 7.20.

As it is shown on the above error matrices, the model meets the original assumptions. The 95% prediction level of the model might have been higher but due to lack of computing power of the device, the model was not subjected to the improvement of hyperparameters. Next the investigations comparing various learning configurations and verifications of the neural networks were carried out involving a few persons tested. 5 persons participated in the test and each of them had to perform a definite facial expression. The test was conducted in various times of the day and room conditions.

The tests were carried out applying the following concepts: (1) Training of the neural network carried out on the data coming from one person, prediction trial on the data of the person whose data was not used during the process of learning; (2) Training of the neural network based on the data coming from all persons tested, prediction trial on the data coming from the same persons but not participating in the process of training; (3) Training of the neural network based on the data coming

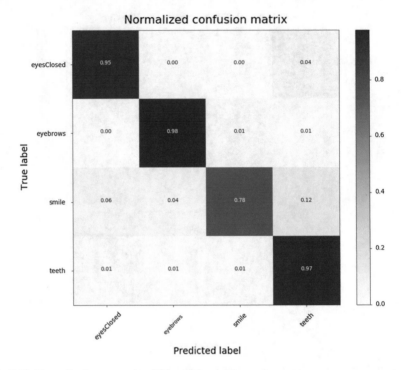

Fig. 7.20 Normalized error matrix of the model

from all the persons tested excluding the person on whom the model performance will be tested.

During the test, the person tested had to perform the following actions: teeth clenching, raising eyebrows, smile, closing eyes. Each type of facial expression performed during the test was evoked every 3 s for 45 s. The whole test together with the change of the record path was carried out within 3 min. Such time caused that the person tested did not experience any discomfort during the test and at the same time could remain properly concentrated. The first investigation consists in verifying whether the model based on the signals from the brain of patient *A* would successfully manage the data coming from person *B*, who did not participate in the process of training. The data distribution is presented in Table 7.5 and is as follows.

After dividing the data into the training data and the test data, error matrices were generated to verify whether the model can manage generalization of the data coming from the same person. The error matrices are shown in Figs. 7.21 and 7.22.

In the next stage, prediction of the model was carried out on the data of the persons who did not participate in the process of training. Using the error matrices shown in Figs. 7.23 and 7.24, generalization of the model trained on person *A* carrying out prediction on the data coming from person *B* is shown.

Based on the error matrices shown in Figs. 7.23 and 7.24 it can be concluded that the model generated based on the data coming from one mind cannot determine

Table 7.5 Data distribution for test number 1

Data	Size
Input data	(7925.14)
Input data—test	(1585.14)
Input data—training	(6340.14)
Output data	(7925.4)
Output data—test	(1585.4)
Output data—training	(6340.4)

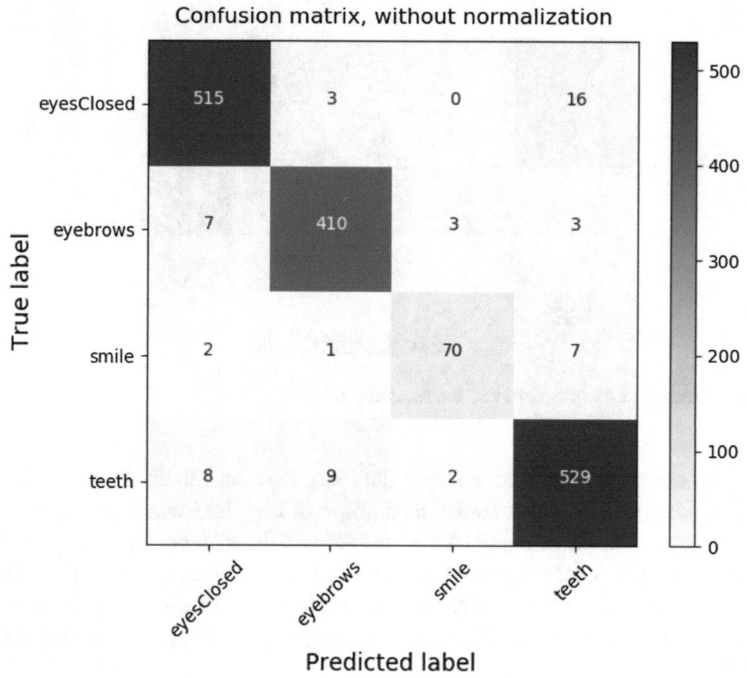

Fig. 7.21 Error matrix of the created model without normalization

dependencies for the mind of another person not participating in the process of learning. Still another prediction was made on the data coming from the other persons but the results are comparable with what was presented in the above-given error matrices. It can be observed that most predictions were wrongly classified as teeth clenching but as the research shows this dependency is accidental.

The aim of the next test was to verify whether an artificial neural network can draw generalizations from electrical signals coming from several persons and perform prediction successfully on the data coming from the same group but never seen by the neural network model in the process of learning. The data distribution based on which the process of training was conducted is shown in Table 7.6.

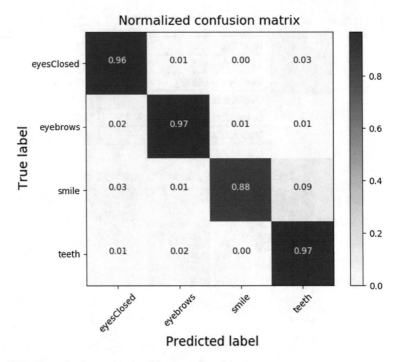

Fig. 7.22 Normalized error matrix of the created model

For the neural network of a larger data structure and their diversification, the model of the neural network shown as a graph in Fig. 7.25 was suggested to begin with. At first the network suggested consisted of 1 input layer, three unseen layers and 1 output layer. These layers contained 14, 10, 8 and 6 neurons, respectively.

Having created the neural network structure for a given case study, the characteristics of learning accuracy and loss were presented, which are shown in Figs. 7.26 and 7.27.

As it is shown in Figs. 7.26 and 7.27, the model is doing very well with data generalization of the *EEG* signals coming from several persons. Prediction accuracy during training and validation is about 93%, which can be considered as a very good result. The loss value during training and validation is almost identical, therefore it may be claimed that the model is not overfitted. In the subsequent stage, error matrices were generated based on the test data of the persons participating in the process of training. The results are shown in Figs. 7.28 and 7.29.

As it is shown on the above-mentioned error matrices, the neural network without any major problems manages generalization of the data coming from several persons and governed by various relationships. Based on this investigation, another investigation was taken up which would check whether, based on the data coming from the set of the persons tested, it is possible to perform correct prediction for the data coming from the person who did not participate in the process of training.

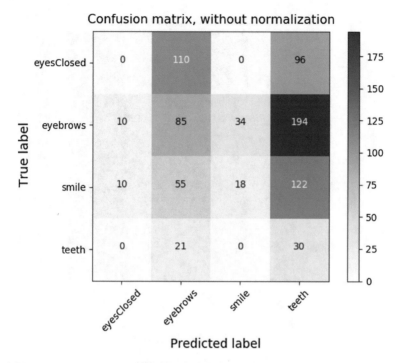

Fig. 7.23 Error matrix of the created model without normalization

The concept of investigation is to verify whether the generalizations generated in the process of training are able to map correctly the signals coming from the brain which did not, at any moment, participate in the process of the neural network learning. The data distribution based on which the process of training was conducted is shown in Table 7.7.

Subsequently, error matrices correlated with the data from the person who did not participate in the process of training were generated. Error matrices, both normalized and without normalization, are shown in Figs. 7.30 and 7.31.

Based on the error matrices shown in Figs. 7.30 and 7.31 it can be concluded that the level of prediction for the person who did not participate in the process of training is about 50%. This coefficient is not high but it is much higher than in the case of training on one person and prediction attempt on the other person. The prediction level on the data presented in test number 3 might be improved by collecting data from o bigger number of persons.

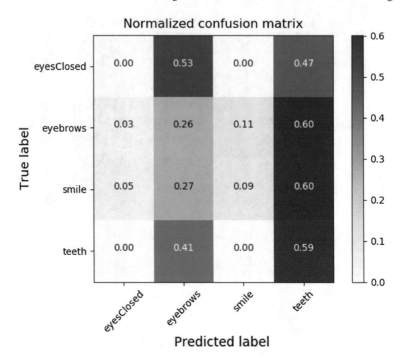

Fig. 7.24 Error matrix of the created model without normalization

Table 7.6 Data distribution for test number 2

Data	Size
Input data	(26,294.14)
Input data—test	(5259.14)
Input data—training	(21,035.14)
Output data	(26,294.4)
Output data—test	(5259.4)
Output data—training	(21,035.4)

Based on the investigation it can be concluded that classification of the *EEG* signal using the neural network trained on one person and verified on the person not participating in the training process is impossible due to various dependencies taking place in human brain during the occurrence of biological artifacts. Nevertheless, the prediction level on the test data coming from the person tested in a single case is very high. This fact can be used for creating the modules of artificial intelligence trained directly on the person who will control an application or a device using biological artifacts. Using neural networks it is possible to perform generalization of the *EEG*

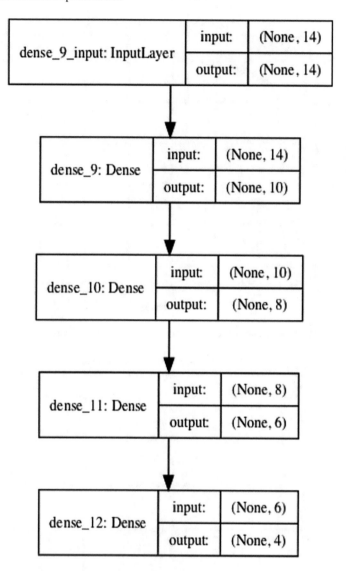

Fig. 7.25 Model of the created neural network showing the layers

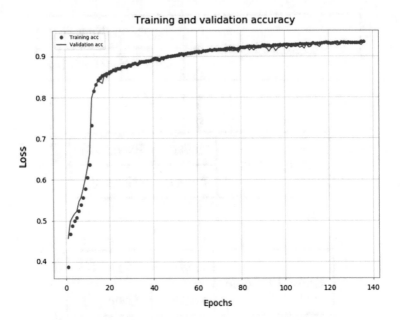

Fig. 7.26 Coefficient of the model accuracy during training and validation

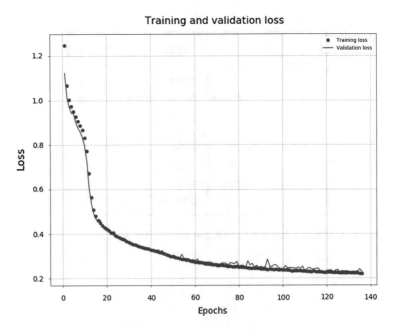

Fig. 7.27 Coefficient of the model loss during training and validation

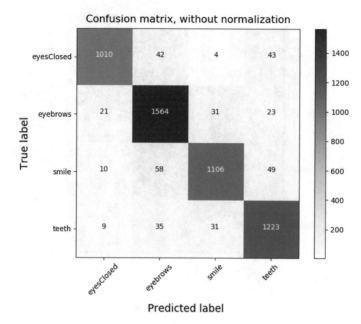

Fig. 7.28 Error matrix of the created model without normalization

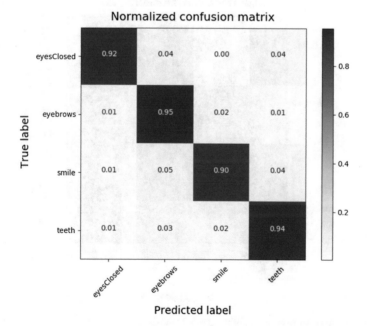

Fig. 7.29 Normalized error matrix of the created model

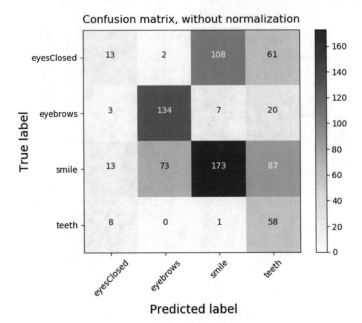

Fig. 7.30 Error matrix of the created model without normalization

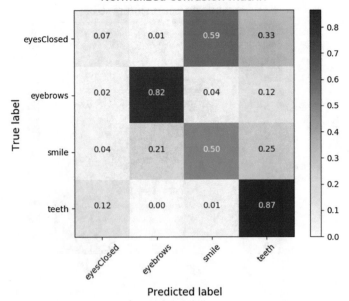

Fig. 7.31 Normalized error matrix of the created model

Table 7.7 Data distribution for test number 3

Data	Size
Input data	(22,074.14)
Input data—test	(4415.14)
Input data—training	(17,659.14)
Output data	(22,074.4)
Output data—test	(4415.4)
Output data—training	(17,659.4)

data coming from many persons in spite of the fact that the brain of each person tested taking part in training behaves in a different way. The prediction level slightly decreases but then the system implemented based on such a neural network might by used by many people. Based on the investigation with index number 3, it was concluded that it is possible to train a model which will be used by the persons never participating in the training process, but certainly it will not be as efficient as the model trained directly on the person using a given software or a device.

Reference

1. Ghaemia, A., Rashedia, E., Mohammad Pourrahimib, A., Kamandara, M., Rahdaric, F.: Automatic channel selection in EEG signals for classification of left or right hand movement in brain computer interfaces using improved binary gravitation search algorithm. Biomed. Signal Process. Control **33**, 109–118 (2017). https://doi.org/10.1016/j.bspc.2016.11.018

Chapter 8
Using BCI Technology for Controlling a Mobile Vehicle

For the purposes of the research carried out, a device Emotiv EPOC+ NeuroHeadset presented in the previous chapters of this monograph was used. After installing the device software, a user receives a wide spectrum of possibilities in the aspect of the EEG signal analysis. The manufacturer provides appropriate software which enables real-time access to the signals from each of the electrodes installed in the device. The device communicates with a computer via a wireless Bluetooth 4.0 adapter. An alternative solution for Emotiv EPOC+ NeuroHeadset may be a custom designed and executed device. When beginning the design works, it was necessary to select a controller which would perform relevant computations, had the option to connect necessary modules and would operate according to a developed algorithm. Raspberry Pi 2 was selected for the purposes of the work. The controller operates on any Linux system. It will act as a computer, computation unit of the entire device, to which all movement commands will be sent. They will be processed with appropriate algorithms and a signal controlling the drive motors and servo-mechanisms will be sent on their basis. Raspbian, normally installed on PI 2, was selected as the operating system. The main algorithm was written in Python. A four-core Raspberry PI 2 processor and 1 GB of RAM memory guarantee computing power reserves. The Raspbian system also enables additional separation of processes into the particular cores of the processor. Thus the operating system may operate stably on one core and the remaining three cores may be used for the needs of other processes. It also has a dedicated output for a camera and the possibility to expand memory. Apart from a big computing power the option to connect additional modules is very important. Raspberry Pi 2 has two independent I2C, SPI, UART buses and 26 programmed input/output GPIO. The controller has 4 USB 2.0 Ports. Therefore it is possible to connect to one of them the Wi-Fi module, a mouse and a computer keyboard, and even an additional disc. Controlling rotation speed of DC motors and steering angle of servomechanisms is performed by the change of the signal duty cycle. In the initial phase of the research program PWM Raspberry PI 2 was used, however, due to high inaccuracy and additional module—Adafruit Mini Kit 16-channel PWM I2C controller—Servo Hat was used. Also TP-Link TL-

WN722N network card with antenna TL-ANT2405CL was used for the work on the device. It features, among others, the possibility to connect a separate antenna. Transmission speed in N to 150 Mbps standard ensures a smooth data transmission, stream video, using internet telephony and online games and 64/128 bit coding WEP, WPA/WPA2/WPA-PSK/WPA2-PSK(TKIP/AES)—compliance with IEEE 802.1X standard. For the purpose of this research, a standard antenna was replaced with TL-ANT2405CL model to increase the range of the module and decrease delays. The gain of 5 dBi omnidirectional antenna makes it possible to cover a larger area with a wireless signal. The operation frequency is 2.4 GHz and is compatible with 11 b/g/n devices. The vehicle is driven by two Lego 8882 motors. Each motor is connected independently and is responsible for rotation speed of one of the rear wheels. The big advantage of the motor is an epicyclic transmission installed inside and a big moment in relation to the motor mass. To increase additionally the motor rotation speed a 1–3 transmission gear was assembled, which enables maximum speed of 600 rpm. Two Tower Pro SG-92 Carbon servomechanisms were used in the designed device. They were installed in such a way as to make each servo change the steering angle of one of two front wheels. In the first phase, the wheel was installed directly on the axle of the servomechanism. The choice of battery was strictly connected with other, formerly selected, sub-assemblies. The number of battery chambers and supply voltage are significant elements. L298N controller module enables the change of direction and rotation speed of both motors. Additional two inputs (A-B Enable) are used to change rotation speed of the motor. Changing values on four inputs Input, it is possible to change the direction of the motor rotation. The big advantage of the system is the 5 V DC output, from which the electronics of the robot may be additionally supplied. In order to ensure a stable operation of Raspberry PI 2 it is necessary to supply it with 5 V DC. The battery voltage is 11.1 V. A step-down converter was used for this purpose. In the first phase Raspberry PI 2 was supplied from the additional output of the bridge H module, however, because of a too high power consumption, an additional converter was applied, which is connected directly between the electronics and the battery. The frame of this four-wheel vehicle is built using Lego Technic modules. Owing to the use of modules, default design of the dimensions and the frame shape is possible. Additionally, through a wide range of blocks, it is possible to modify the frame of the device during operation. Another advantage is the gears, which can be used to form a transmission. The devices also uses motors, which perfectly fit the frame, forming together a stable structure. The 62.4 × 20 cm slick wheels used ensure excellent grip. A complete device weighs about 1 kg. The construction of the frame started with building a transmission from the motors onto the wheels. Next, the modelling servos were remade in such a way so as to fit the frame. In the first stage, the wheels were steered directly by the servo. Unfortunately, a disadvantage of this solution is the fact that a travelling wheel tilts and transfers the vibrations onto the servomechanism, which causes vibrations and accelerates the wear of the servo. The problem was solved by installing a 1–3 transmission gear. The Raspberry Pi 2 controller together with additional modules was mounted in the centre of the vehicle. The converter is in the gap between the motors. The batteries are placed at the rear. It is the heaviest sub-assembly of the four-wheeler. Installing it on the drive axle is to

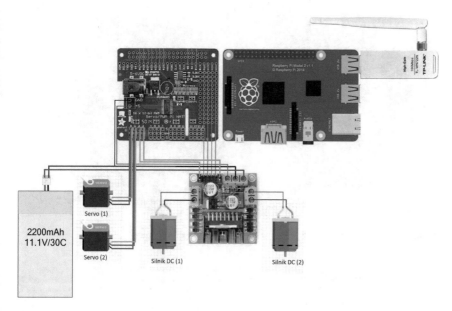

Fig. 8.1 Electrical diagram of the system used in the research investigations carried out

ensure additional downforce. The H-bridge was installed at the top. Figure 8.1 shows the electrical diagram of the device. Since the Servo Hat extension plugged into a Raspberry Pi 2 occupies all additional inputs, the module (H-bridge) was plugged into the exit from the second plate. The system is powered by a 11.1 V DC 2200 mAh lithium-polymer battery. A Wi-Fi module is used to program the system and for further transmission to set appropriate movement parameters. Electrical motors were used as a drive in the vehicle and the servomechanisms for steering. Since the motors operate at maximum 9 V DC voltage and the system is powered with 11.1 V DC, the PWM signal maximum fill was limited, which translates into output voltage. This way, the H-bridge will never receive a higher fill, which would mean voltage higher than 9 V DC. The PWM signal maximum fill value was measured before the motor was connected to the system.

Controlling the devices requires the creation of a network in which all the necessary components will be able to cooperate. The Bluetooth and Wi-Fi technologies were selected for the purposes of this work. All the parameters downloaded from the Emotiv EPOC+ NeuroHeadset are sent to a PC class computer via Bluetooth 4.0. The computer operates in Wi-Fi network to which a four-wheel mobile robot was also connected [1]. By using the SSH protocol, a network user has access to a virtual console of the device and is able to send appropriate commands this way. The control algorithm was divided into two parts. The first one was implemented in the robot and it consists of the mobile abilities of the devices, which means they specify what the device should be doing receiving a given command. The second part of the

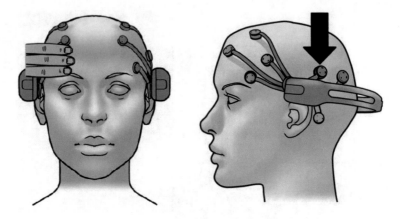

Fig. 8.2 Proper placement of the Emotiv EPOC+ NeuroHeadset device

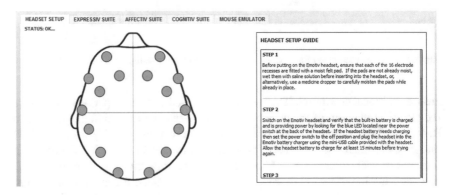

Fig. 8.3 EPOC Control Panel with proper placement of the device

algorithm was implemented on a PC. In this case, all commands are generated on the basis of the pulses collected from the Emotiv EPOC+ NeuroHeadset device and sent to the mobile robot. The work with the Emotiv EPOC+ NeuroHeadset device was commenced with soaking the electrodes with—the felt items put on them—with saline solution. The measurement accuracy depends on both the correct placement of an electrode and its contact with the skin. After soaking the electrodes, it was necessary to put the device on the head as shown in Fig. 8.2.

After starting the EPOC Control Panel program and communicating with the device, the software interface signalled if the electrodes were properly correlated with the skin. The system itself needed 1–2 min to detect and properly identify the electrodes. A properly configured device is shown in Fig. 8.3.

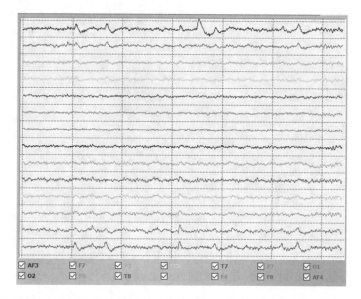

Fig. 8.4 An EEG signal for the condition of standard-normal activity of a tested person on 14 transmission channels

After establishing a connection and proper location of electrodes, the signals were identified based on the potentials generated by facial expressions. The software provided with the devices enables the view of a virtual face (avatar) of the controlling person. Figure 8.4 shows the condition of standard activity when the person tested does not perform any specific actions.

Furthermore, Fig. 8.5 shows the EEG signal changes at the eye lids movement. It can be observed that electrodes AF3, AF4, F7 and F8 registered a definite change of the signal amplitude. In Fig. 8.5 it can be also counted that the person tested performed 13 cycles of eye blinking. Figure 8.6 shows the EEG signal when the person tested clenched the teeth. It is clearly seen that all the electrodes registered a considerable change of the EEG signal. This is because when a person tightens a jaw, most head muscles are tightened. When the muscles are tightened they become to vibrate and also generate a pulse, which was registered by the device. Figure 8.7 also presents an EEG signal adequate for eyebrow movement. It can be also noticed that only some electrodes recorded certain signal amplitude changes.

The vehicle was controlled through detecting appropriate, certain behaviours observed in the EEG signal—artifacts, induced by facial expressions. Classified artifacts and their assigned movements are presented in Table 8.1.

Selecting an appropriate sensitivity of the measured signal in relation to the one generated for the four-wheel vehicle is important for such control. Too frequent blinking or an intensive smile may cause lack of comfort noticeable during the controlling process. The data receiving software was developed in such a way so that

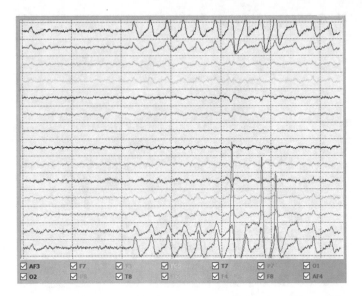

Fig. 8.5 An EEG signal for the condition of eye blinking of a tested person on 14 transmission channels

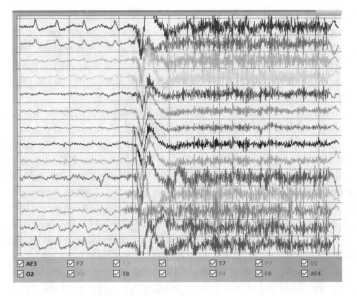

Fig. 8.6 An EEG signal for teeth clenching of a tested person on 14 transmission channels

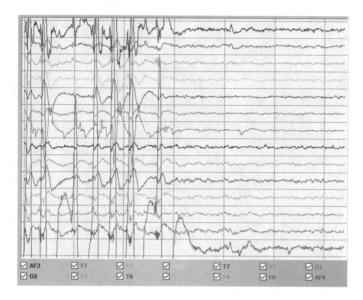

Fig. 8.7 An EEG signal for eyebrow movement of a tested person on 14 transmission channels

Table 8.1 Classification of received signals in relation to the command generated	Classified artifact	Movement of a four-wheeled robot	Assigned sign
	Blinking eyes	Driving forward	„w"
	Blink of the right eye	Turn right	„d"
	Blink of the left eye	Turn left	„a"
	A clamp of teeth	Stop	„"
	Smile	Driving backwards	„s"
		Exit	„r"

it is not necessary to maintain a characteristic artifact, in order to maintain the required movement. Unfortunately, the Putty software granting access to a virtual console requires confirming each command with Enter [2]. The code program including the main control loop used in this implementation is shown in Appendix.

References

1. Moreno, R.J., Aleman, J.R.: Control of a mobile robot through brain computer interface. INGE CUC **11**(2), 74–83 (2015)
2. Paszkiel, S: Using the Raspberry PI2 module and the brain-computer technology for controlling a mobile vehicle. In: Szewczyk, R., Zieliński, C., Kaliczyńska, M. (eds.) Progress in Automation. Robotics and Measurement Techniques. Advances in Intelligent Systems and Computing, pp. 356–366. Springer, Switzerland (2020). https://doi.org/10.1007/978-3-030-13273-6_34

Chapter 9
Using BCI Technology for Controlling a Mobile Vehicle Using LabVIEW Environment

Emotiv EPOC+ NeuroHeadset was used for the purpose of the research experiment. Its advantages are a high resolution of data obtained in a short time, a relatively quick calibration and a very good value for money. The downside of this device is noticeable time delays in signal transmission to the workstation. EPOC Control Panel is the first program that is run after the Emotiv EPOC+ NeuroHeadset is set up. After the appliance is placed on the head, it must be checked whether all electrodes touch the skin properly.

A robot described in the previous chapter was used in order to physically execute the control process. The choice of LabView program environment is justified by a very small number of available solutions in the scope of correlation of LabView with BCI technology. Moreover, the LabVIEW environment has many built-in features used to communicate with multiple devices. Another advantage of LabVIEW is a wide range of functions for data visualisation. In order to build the program, a mechanism for joining icons that are available in the environment was used. A sample EEG signal reading from 14 electrodes in the LabVIEW program is shown in Fig. 9.1.

For communication of Emotiv EPOC+ Neuroheadset with the computer, the Lab-VIEW Emotiv Toolkit V2 package was implemented. The package only works with a 32-bit version of LabVIEW and requires the user to show the exact location of the original edk.dll file. The package allowed for the use of all signals implemented by Emotiv Inc. In this way it is possible to verify whether facial expression readings are correct or not by simultaneously watching them in the EPOC Control Panel program. A virtual HMI operator panel where LEDs were placed (Fig. 9.2) was created in order to make sure that the LabVIEW program reads signals correctly. A smile or another previously identified facial expression generated a light pulse in a corresponding LED controls. The available package returns True or False values by default (Fig. 9.3). However, in the case of clenching teeth the package returns a number, therefore for clenching the value above zero was replaced with True.

© Springer Nature Switzerland AG 2020
S. Paszkiel, *Analysis and Classification of EEG Signals for Brain–Computer Interfaces*, Studies in Computational Intelligence 852,
https://doi.org/10.1007/978-3-030-30581-9_9

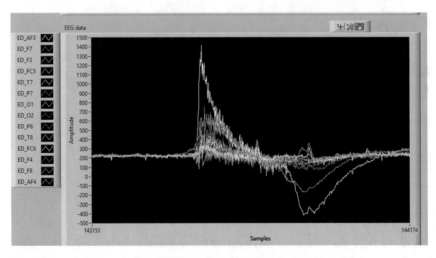

Fig. 9.1 Sample EEG signal reading from 14 electrodes in the LabVIEW program

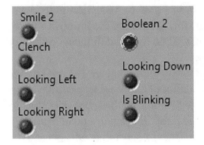

Fig. 9.2 LED controls used to detect a given signal model

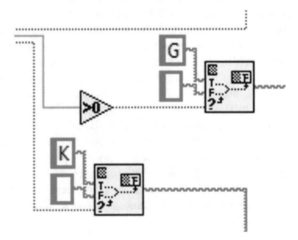

Fig. 9.3 Changing true signal to a given robot action sign according to Table 9.1

Table 9.1 Signs corresponding to actions/movement directions of the robot

Action/direction of movement	Sign
In front	G
Backwards	T
Stop	S
Horn	K
To the right	P
To the left	L
Reset	R

The research experiment was conducted based on the Emotiv EPOC+ Neuro-Headset. Emotiv EPOC+ differs from the EPOC version, among others, in a higher frequency sampling rate of the signal measured. For the purpose of the task, Lab-VIEW was chosen, which is based on G programming language and one of the niche development environments for such purposes so far. The electroencephalographic signal was obtained owing to the research licence of the Laboratory of Neuroinformatics and Decision-Making Systems of Opole University of Technology for Emotiv hardware and software containing Emotiv SDK with access to data in a *.raw format. Software Development Kit is a set of developer tools vital for writing applications that use a given Emotiv Inc. library. For the purpose of the control process execution, the signs corresponding to specific actions—movement directions of the robot were defined—which is shown in Table 9.1.

In the first testing phase, the connection with the mobile robot was established via Bluetooth pairing with the computer on MS Windows 10. Then the port was chosen in the LabVIEW environment, e.g. COM7. The disadvantage of such a solution is the necessity of pairing the robot with MS Windows every time in order to establish the connection. After the port is chosen in the panel, any sign can be sent in order to check whether the communication with the mobile robot is properly executed.

The HMI panel shown in Fig. 9.4 consists of the following modules: String to write—test signs sent to the mobile robot; Resulting string—signs received or testing whether signs are sent; X Value—current gyroscope value on X axis (horizontal); STOP—program stopping; Boolean—emergency switch, it resets the signs received in the mobile robot; Cognitive Action—currently operating learned signal, the same as in the case of controlling a cube in EPOC Control Panel; Sample Count—graph showing proper receiving of samples from the electroencephalograph; EEG Data—graph showing EEG signal readings from fourteen electrodes; VISA resource name—COM port of the synchronized robot.

In Fig. 9.5 the following work parameters of the program are defined: baud rate 9600, 8 bits of data, and 250 ms delay—¼ of a second, so sending was at a frequency of 1 Hz for the C2000 microprocessor. In order to control the mobile robot on the basis of potentials evoked in Table 9.2, facial expressions were assigned to the actions/movements directions described above. This table correlates with Table 9.1 as far as the first column is concerned (Fig. 9.6).

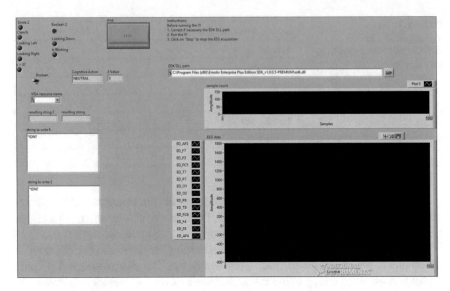

Fig. 9.4   Virtual HMI panel

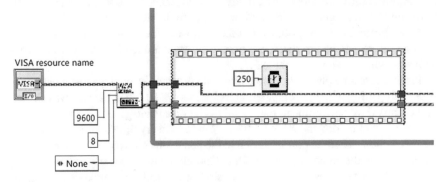

Fig. 9.5   Part of the program in LabVIEW responsible for communication

	Action/direction of movement	Facial expressions
Tab. 9.2 Facial expressions corresponding to the actions/movement directions of the robot	In front	A clamp of teeth
	Backwards	Look down
	Stop	Smile
	Horn	Blinking eyes
	To the right	Look to the right
	To the left	Look to the left
	Reset	Emergency switch on the console

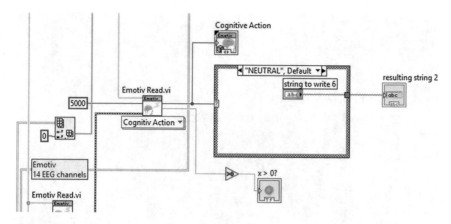

Fig. 9.6   Attempt to use learned signal in the LabVIEW environment

The tests of evoked potentials were the basis for creating the control system. The experiment was based on finding such signals that the robot operator identifies at an efficiency of 100%. To that end, EPOC Control Panel was used. As shown in the experiment conducted, expressions such as smile, clench, blinking were executed at a high efficiency and high interference rejection. Tests that involved rotating a solid— the cube—were much more difficult to execute because its signal identification and classification require maximum focus. A sample change of the EEG signal amplitude versus time for blinking fivefold is also shown in the LabVIEW graph (Fig. 9.7).

The characteristics of the EEG signal amplitude change on particular transmission channels is shown in Fig. 9.8, whereas Figs. 9.9 and 9.10 show the EEG signal amplitudes versus time for the robot operator looking to the left (Fig. 9.9) and to the right (Fig. 9.10).

Figure 9.11 shows the EEG signal amplitude change in the LabVIEW environment while smiling.

The EEG signal amplitudes presented show differences in recognizing the particular signals, for example the electroencephalogram for looking to the right (Fig. 9.10) differs considerably from the electroencephalogram of looking to the left (Fig. 9.9.), in which case two additional electrodes receive the electric pulse present while looking to the right. The LabVIEW environment shows EEG reading on a different scale; therefore, the signal corresponding to smiling is noticeable more clearly (Fig. 9.11). Clenching generates a strong signal on all channels (Fig. 9.8). Blinking either the left or the right eye is recognisable as in the case of the direction of looking. The EEG blinking signal is reflected best because on its basis it can be easily identified how many times a given person has blinked (Fig. 9.7).

Based on literature studies conducted, it can be concluded that the number of similar executions of the use of the Emotive Inc. hardware potential in correlation with LabVIEW is very small. What is more, as far as the analysis and classification of EEG artifacts executed for the purpose of the robot control using Emotiv EPOC+ Neuroheadset with LabVIEW are concerned, such implementation attempts are some

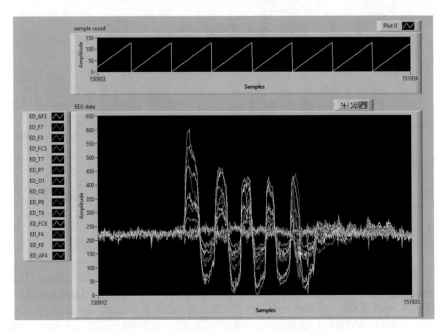

Fig. 9.7   EEG signal visualisation while blinking fivefold in LabVIEW

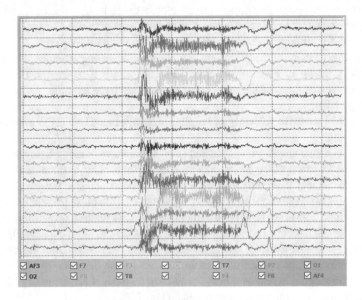

Fig. 9.8   EEG signal amplitude change while clenching

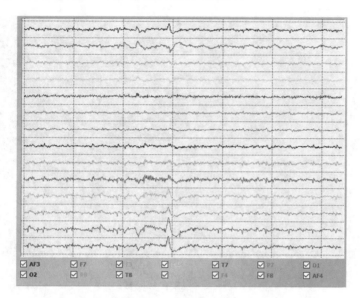

Fig. 9.9 EEG signal amplitude change while looking to the left

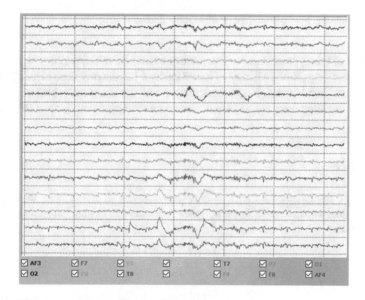

Fig. 9.10 EEG signal amplitude change while looking to the right

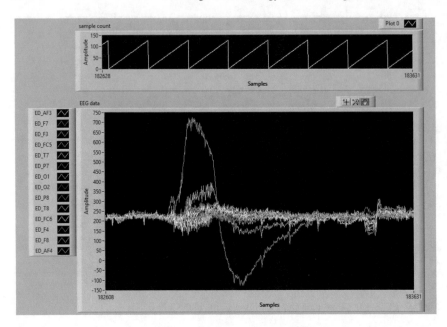

Fig. 9.11 EEG signal amplitude change while smiling in LabVIEW

of the first of this type in Poland. The control of mobile vehicles using older Emotiv Inc. hardware (e.g. Emotiv EPOC+ NeuroHeadset) has already been researched both in Poland and in the world, but instead independent applications written in JAVA, C# programming languages etc. [1] were used in most implementations of this kind.

Reference

1. Paszkiel, S.: Facial expressions as an artifact in EEG signal used in the process of controlling a mobile robot with LabVIEW. Electr. Rev. **93**(4), 156–160 (2017). https://doi.org/10.15199/48. 2017.04.38

Chapter 10
Augmented Reality (AR) Technology in Correlation with Brain–Computer Interface Technology

Immersion allows you to "absorb" the human being into the electronic reality. Then you have to deal with the so-called immersion of the senses. Immersion can be seen in many cases, including the area of arts, culture and technological environment. In our case the latest range of immersion is the most important from the point of view of the technology discussed. Immersion in this range is connected directly with the electronics of advanced immersive technologies. Cave Automatic Virtual Environment and Head Mounted Displays are examples of immersive technologies already developed. Cave Automatic Virtual Environment (CAVE) creates the environment by organizing an enclosed space in the area perceived by reality. The technology consist in projecting digital images on the walls, at the same time it is based on the three-dimensionality. Head Mounted Displays (HMD) is a technology which allows a man enter the world of virtual reality owing to a device attached to the head in the shape of a helmet and a suit holding sensors. The main difference between the CAVE and HMD technology is that in the space of the CAVE technology a recipient is surrounded by environment, and the system recipient of the HMD technology is in his own space created in a digital way.

In science, there exists the concept of reality known as Augmented reality (AR), which combines virtual and real world. Under this system, with the help of special glasses we are able to observe the world around us in the streets of the cities and the elements produced by virtual reality. The assumptions, on the basis of which Augmented reality is based on, is the combination of the two worlds mentioned before, a real-time interaction and freedom of movement in three dimensions. It should be noticed that, among others, at the Massachusetts Institute of Technology (MIT) more and more lectures for the students are conducted on the basis of Augmented reality. Students use their smart phones and GPS devices gradually exploring previously enriched campus sites for learning. Augmented reality can be used in the study to obtain information on the objects on which students work, as is the case at MIT in the U.S.A., by immediate verification and electronic feedback in the form of information; in medicine—an access to the data on the structure of the internal organs of

© Springer Nature Switzerland AG 2020
S. Paszkiel, *Analysis and Classification of EEG Signals for Brain–Computer
Interfaces*, Studies in Computational Intelligence 852,
https://doi.org/10.1007/978-3-030-30581-9_10

a patient; in marketing, and the most significant area in terms of this monograph in robotics—by identifying objects [1] among which the robot moves and supporting the generation of potentials evoked in the brain of a person who controls the robot motion by using BCI technology in the feedback loop.

NeuroSky is a manufacturer of devices which are technologically based on the concept of brain–computer interfaces. Their wide range of products includes among others MindWave Mobile, which is a device shaped like headphones with a ground electrode attached to the ears [4]. This device is based on the electroencephalographic examination. The EEG signal is measured during its work and thus activities of individual brain waves are observed. Based on the above facts we can execute the process of control and influence in different ways different objects in space, including the movement of a mobile robot.

When testing the device in the Laboratory of the Institute of Automation at Opole University of Technology, it was possible to use it in order to evoke intentional activities in terms of the control process. Stimulation of the brain, by driving it in certain states of meditation and enhanced thinking, brought practical results. An identification of changes in the amplitude of the EEG signal in correlation with induced potentials enabled the use of the product to carry out simple tasks. Figure 10.1 shows a visualization of brain wave activities identified by the application available with MindWave Mobile.

Google Glass Enterprise Edition of a module structure was created based on its predecessor Google Glass and is a typical example of a device based on partial immersion, namely augmented reality. Owing to certain constructional solutions, they can operate as devices supplementing the control process based on the technology of brain–computer interfaces. The Google Glass Enterprise Edition operates on

Fig. 10.1 A window showing the visualization of brain wave activities identified by the applications available with MindWave Mobile by NeuroSky

the Android operating system. It can combine its function with corrective glasses used by people with vision defects, therefore they do not discriminate this group of people in the context of lack of possibility of using this technology in practice. The implementation the Android operating system in the device by Google will enable free access to many free applications and definitely facilitate the operation of combining BCI technology with the solutions of augmented reality. The resolution of the images displayed by Google Glass Enterprise Edition will be definitely sufficient for using a display as an element supporting the control process using BCI technology. The Google Glass Enterprise Edition device is equipped with 2 GB of operational memory RAM and min. 16 GB of bulk memory, which meets the requirements of the suggested system solution presented in the further part of this chapter. A built-in USB port enables charging the battery in the device.

Google, in its Google Glass device, implemented the mechanism of voice transmission to the person using the glasses based on bone conduction. A sound wave is transmitted via a built-in microphone located on the temporal bone of a person directly to the middle ear. Unfortunately, this might affect the process of the EEG signal acquisition, therefore the system suggested will not use this device functionality. Another feature of Google Glass is the fact that the device is controlled by voice and touch sensing but in the case of the system neither of these control methods will be needed. The application of the brain–computer interface will enable the use of the device for control processes without the need to control it.

As regards executing of the control process, the issue of the power supply for the device becomes significant in a longer uninterrupted time. There already exist Google Glass models with bigger capacity batteries and thus they can operate for longer periods of time. Controlling with human thoughts without using evoked potentials is presently difficult to implement in everyday conditions, which was proved during author's neuromarketing research. Nevertheless, controlling based on evoked potentials is relatively easy to implement. In order to avoid manipulating with the limbs or closing the eyes, using voice commands, we can, using Google Glass Enterprise Edition, emit images on glass and thus stimulate our brain in the context of inducing intended changes in the amplitude of brain waves. Based on this, in the next stage we are able to control both the application on the workstation and the mobile robots exploring a given area using the correlation of the brain–computer interface and the Google Glass Enterprise Edition glasses. Obviously, the images emitted on glass would be coupled directly with what the eye of the controlling person can see and their relevant formula would have a positive impact of stimulating the brain in the context of the control process. Table 10.1 shows sample correlations between the human brain state, activity of the particular types of brain waves, presented images and a planned action connected with controlling a mobile robot [7].

The tests connected with controlling a Quadro Helicopter using Google Glass Enterprise Edition have been carried out since Google built the device. The camera installed in Google Glass Enterprise Edition identified the head position of the person controlling the Quadro Helicopter [3] in relation to the reference point, which, in this case, was the floor. Head movements had impact on the process of moving with Quadro Helicopter [2, 6]. It should also be taken into account that the Google

Glass technology found its application in medicine. Here we can mention the case of a doctor operating a patient using Google Glass to carry out a videoconference via a camera installed in Google Glass and providing information (pertaining to the course of operation) to his interns. It was possible owing to the use of the tool Google Hangouts. While developing a system and carrying out the analysis for and against using the Google device, the problem of wind which can cause noise in the loudspeakers and the microphone was encountered. Nevertheless, the system offered does not use the audial part of the device, i.e. functions without a loudspeaker or a microphone thus this type of interference would not have any impact on the operation of this system.

Figure 10.2 shows a general architecture of the system operation based on Augmented Reality and brain–computer Interface technology. The brain and eyes of the controlling person are the two units without which the system offered will not work. Its advantage is the fact that both voice signals and the need for the use of upper and lower limbs are not necessary. The system is based on Augmented Reality in which

Table 10.1 Sample connections between the human brain state, activity of brain waves, presented image and an action planned

The condition of the human brain	Activity of brain waves	Presented picture	Planned action
Meditation	Alpha waves	Forest, sea	Stop moving robot
Increased activity	Gamma waves	Problem to solve—a task	Robot movement Forward

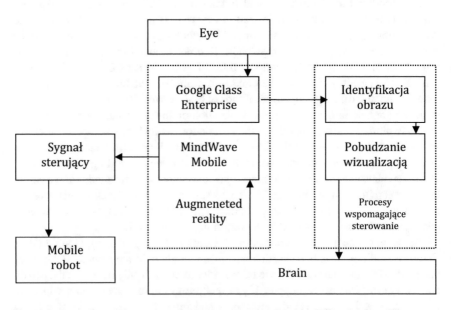

Fig. 10.2   An outline of the concept of using augmented reality with BCI technology [5]

the main role is played by Google Glass Enterprise Edition. A block of the system known as Processes which support controlling consists of: identification of an image observed by the user of the system and stimulation based on the visualization of images directly related to the observed reality. A control signal is identified for the mobile robot control processes after producing the appropriate response in the brain by means of MindWave Mobile by NeuroSky.

Currently, there are practical implementations of technology based on augmented reality among other things in the entertainment industry for the construction of urban games. Augmented reality can also be used among other things to scan products at shelves in a store, for an immediate comparison of their composition, caloric testing, prices etc. The development work were carried out also in the context of the People+ or the face identification of people with whom interviews are conducted and to obtain information about them. The field to use Augmented reality is wider and wider. BCI technology may become in the future a great addition to the above-mentioned technologies [5].

References

1. Bielińska, E., et al.: Identification of Processes. Silesian University of Technology, Gliwice (1997)
2. LaFleur, K., Cassady, K., Doud, A., Shades, K., Rogin, E., He, B.: Quadcopter control in three-dimensional space using a noninvasive motor imagery-based brain-computer interface. J. Neural Eng. **10**(4), 046003 (2013)
3. Paszkiel, S., Sikora, M.: The use of brain-computer interface to control unmanned aerial vehicle, automation 2019, progress in automation, robotics and measurement techniques. In: Szewczyk, R., Zieliński, C., Kaliczyńska, M. (eds.) Advances in Intelligent Systems and Computing, Conference Proceedings Citation Index (ISI Proceedings), pp. 583–598. Springer, Switzerland (2020). https://doi.org/10.1007/978-3-030-13273-6_54
4. Paszkiel, S.: Akwizycja sygnału EEG przy pomocy NeuroSky MindWave Mobile na potrzeby procesów sterowania realizowanych z poziomu systemu Android. Poznań Univ. Technol. Acad. J. Electr. Eng. **84**, 237–244 (2005)
5. Paszkiel, S.: Augmented reality of technological environment in correlation with brain computer interfaces for control processes. In: Advances in Intelligent Systems and Computing 267—AISC. Springer, Switzerland, pp. 197–203 (2014)
6. University of Florida: https://www.eng.ufl.edu/newengineer/news/mind-controlled-drones-race-to-the-future. Jason Dearen (2017)
7. Voznenko, T.I., Chepin, E.V., Urvanov, G.A.: The control system based on extended BCI for a robotic wheelchair. Procedia Comput. Sci. **123**, 522–527 (2018). https://doi.org/10.1016/j.procs.2018.01.079

Chapter 11
Using BCI and VR Technology in Neurogaming

The first remarks pertaining to neurogaming appeared at the beginning on the 21st century. The first game 'Brain Age Nitendo' using the brain–computer interface appeared in 2005. The first games which appeared on the market and used this technology were the games of the 'brain training' type and with time they evolved into more advanced 'brain training' applications from the companies such as, e.g. Lumos Labs. Shortly thereafter, the games using motion detectors such as: Wii Fit Nintendo, Microsoft Kinect Sports and Intel Perceptual Computing appeared. New startups had problems with selling their own devices using the brain–computer interface in neurogaming as their promotion was very expensive. Only about 2010 NeuroSky, Emotiv Inc. and some other companies managed to become known in this field and encourage the potential customers to purchase and use these devices. Brain–computer interfaces, due to correlation with an external stimulus or lack of it, may be divided into asynchronous—based on spontaneous brain activity not related to an external stimulus, and synchronous—related to the occurrence of an external stimulus. The conducted literature review indicates that the development of brain–computer technology determines the creation of a constantly increasing number of software solutions for signal analysis and identification, as well as prototype products for showing changes of the brain electric activity using LEDs.

Brain–computer technology in a non-invasive version gains more and more possible practical applications in different domains of life, including neurogaming. The main advantage of this technology is the possibility to affect video game action using only brain signals without using them directly to trigger the effector muscle of a given limb, which is the case in standard controlling. As shown by the literature review, one of the first neurogames was NeuroRacer, aiming to develop cognitive abilities. Studies conducted using the game allowed to draw a conclusion that people playing the game significantly improved their working memory and cognitive skills. An important positive aspect from the game was also enhancing the multitasking capacity for mental operations, a decline of which is particularly observed in the elderly.

© Springer Nature Switzerland AG 2020
S. Paszkiel, *Analysis and Classification of EEG Signals for Brain–Computer Interfaces*, Studies in Computational Intelligence 852,
https://doi.org/10.1007/978-3-030-30581-9_11

The performed analyses indicate that neurogaming has a relatively low competitive level due to the presence of large delays between a signal transmitted from the surface of a human head to a workstation, and also due to a large number of artifacts accompanying the measurements, which are hard to eliminate in practice. This requires the use of proper signal filtering. One of them is FFT, i.e. a Fast Fourier Transform. This involves a transforming process which gives a transform as a result. The Fourier Transform determined for a discrete signal, i.e. a signal based on a specific number of samples is called a Discrete Fourier Transform. If $x_0, \ldots, x_{N-1}$ are complex numbers, then the DFT is expressed with the formula (11.1).

$$X_K = \sum_{n=0}^{N-1} x_n e^{\frac{-2\pi i}{N} nk}, \quad k = 0, \ldots, N-1 \tag{11.1}$$

One of the components of the Fourier series is a harmonic component that represents the signal in the spectral domain. By using the DFT, signal samples $a_0, a_1, a_2, a_3, \ldots, a_{N-1}$, assuming $a_i \in R$ are transformed into series of harmonics $A_0, A_1, A_2, A_3, \ldots, A_{N-1}$, assuming that $A_i \in C$, it is done according to the Eq. (11.2) where i—imaginary unit, k—harmonic number, n—signal sample number, a_n—signal sample value, N—number of samples.

$$A_K = \sum_{n=0}^{N-1} a_n w_N^{-kn}, \quad 0 \le k \le N-1$$

$$w_N = e^{i\frac{2\pi}{N}} \tag{11.2}$$

When conducting the FFT analysis it should be noted that biological signals are never a sinusoidal signal, but rather a component of many. The FFT analysis enables a quick and accurate identification of signal components. Figure 11.1 shows an example of the FFT analysis conducted using Emotiv TestBench. The data were taken from O_2 electrode for maximum signal amplitude in the range from 80 to -60 dB.

Figure 11.1. shows a signal as gain and frequency, while on the right hand side there are classifications of individual rhythms of brain waves for a signal in gain frequency plane. A high activity of alpha waves in the range from 7 to 13 Hz is

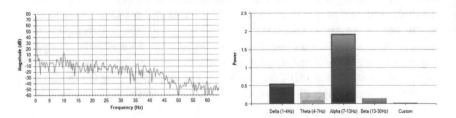

Fig. 11.1 FFT signal analysis carried out using the Emotiv TestBench application

clearly visible here, which is related to idle/relaxed state of the person tested. For comparison, the range of beta waves is very low in this case, which is related to no information processing in a given time window. Thus it can be concluded that the FFT analysis provides a correct signal filtration and identification of activity of given brain wave ranges for specific electrodes for the neurogaming purposes. A dynamic development of games also implies the need of increasing the number of signals recognised by brain–computer interfaces.

The studies conducted at the Laboratory of Neuroinformatics and Decision Systems of the Opole University of Technology, 14 channel electrodes + 2 reference electrodes contained in the Emotiv EPOC+ NeuroHeadset device were used. This device does not require the use of gel or conductive paste, however, saline solution is helpful in its correct operation by saturating felt elements at skin-electrode contact place. An important factor in ensuring correct device operation is its proper location on the scalp of the person tested. Device speed and accuracy is ensured by sampling 2048 times per second, and sample filtration. Once the sampling process has been completed, 256 samples per each channel are obtained of maximum resolution of 16 bits. The device performs FFT filtration. The frequency to which the device responds is in range from 0.16 to 43 Hz, which makes it possible, among others, to read the state of: meditation, idleness, boredom and focus.

The electrodes are located on the head of the controlling person based on international 10-20 standard, as shown in Fig. 11.2. It should be noted that CMS and DRL are reference electrode and grounding electrode, respectively. Other electrodes following the standard of the International Federation for Clinical Neurophysiology have in their elements even numerical identifiers for the right hemisphere and odd ones for the left hemisphere, as well as letter identifiers based on brain lobes over which they are located.

In terms of controlling avatars in a virtual reality, there is still a problem with the BCI today as the device user calibration is a very time-consuming process that must be performed prior the commencement of controlling. In order to identify control signals after correct placing the device on the head of the person tested, it is necessary to run EPOC Control Panel application to verify the connection status and thus proper correlation at skin-electrode interface for all 16 electrodes presented in the visualization. EPOC control panel should be also used for performing cognitive training, the effect of which can be used in neurogaming applications. Cognitive training involves archiving brain activity in a given time in correlation to a specific stimulus, and then comparing archived patterns with observed events. For the performed tests, it was necessary to identify brain activity in a relaxed state and increased mental effort aimed at movement in a specific direction (north, south, east, west) shown in Fig. 11.3—on the right side, marked by a cube in 3D space.

Virtual reality originated in the 1960s [4]. However, recently its development became much more dynamic due to the appearance of new technological solutions such as the goggles that can display pictures both in 2D and 3D mode. Such goggles have two miniature screens, each of them showing a proper part of the image. From the practical perspective, virtual reality can be defined as a combination of special equipment and software. Software solutions play a role of supporting hardware accel-

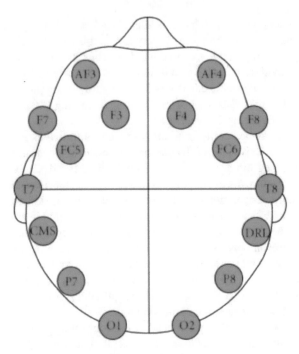

Fig. 11.2 Location of electrodes on the head of the person tested in accordance with 10-20 standard for Emotiv EPOC+ NeuroHeadset together with identifiers

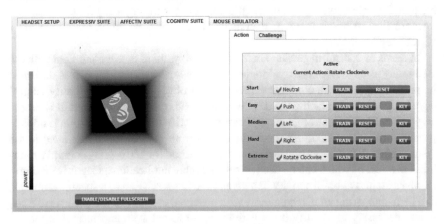

Fig. 11.3 Outline of the head in the main window of EPOC Control Panel and cognitive training performed in Cognitive Suite

Fig. 11.4 Main window of EmoKey application and an example of virtual environment Spirit Mountain created based on Unity engine

erators in the scope of transforming environment into image, which implies a large amount of mathematical computations. However, hardware solutions may support immersion, i.e. dive deeper into generated virtual reality. Currently, visualisation of virtual reality is based mainly on the implementations used in the devices produced by Oculus. Unfortunately, as shown by experiments so far, this technology does not currently allow experiencing virtual reality for longer periods of time, as it is fatiguing to human body, particularly to eyeballs.

Unfortunately, present-day use of BCI for neurogaming requires aforementioned concept of controlling using additional controllers. This is mainly due to the difficulty in distinguishing information from electroencephalography tests for controlling game action, based on several different mental states. However, it is worth noting that for such neourogaming tools as Spirit Mountain (Fig. 11.4) by Emotiv Inc., the application of BCI technology proved to work very well in practice, which was confirmed by the tests conducted in the Laboratory of Neuroinformatics and Decision System of the Opole University of Technology. This is caused by the fact that the number of handled actions is rather low. In a virtual world modelled in this application, it is possible to lift objects by thinking. Moreover, one can move objects in relation to each other by focusing his or her attention. The game is based on the world exploration using internal imaginations based on the induced potentials that are additionally supported by a gyro, which is a part of the equipment of Emotiv EPOC+ NeuroHeadset. This is because the manufacturer wishes to improve the sense of the first-person game-play, and to eliminate mechanical components such as a computer mouse from the control process.

For the purpose of carrying out the tests of neurogaming applications based on brain–computer technology which is used by Emotiv EPOC+ NeuroHeadset, it was necessary to correlate buttons with some specific human mental states as shown in Fig. 11.4. The tests carried out indicate that controlling virtual objects using the brain is a complex and difficult process. Additionally, there is a delay during information transfer, which has a negative impact on the control process. Cerebral Constructor was another tested neurogaming application involving Emotiv EPOC+ NeuroHeadset for the purposes of this experiment. In this case, controlling allowed identification of seven brain activity states associated with moving up, down, left and right, lifting, pulling and lowering used for rotating a given object. As shown by the performed tests, from the practical perspective, there are only three commands required: one for rotation and two for moving. In the case of applying BCI technologies for games other than neurogaming ones, there are some problems arising, mainly related to lack of direct correlation between the products.

Another important problem in terms of virtual reality operation is lack of model cooperation of human body with the technique. In the case of the sense of sight there is an identified issue of lagging—a delay that occurs between the head movements registered by VR goggles and an image generated on the display of the workstation. Still another problem is related to the anatomical labyrinth and its erroneous identification of orientation in relation to gravity as compared to the orientation calculated by the algorithm operating in a given application. A significant issue is also the fact that image is created by VR goggles at fixed distance from the participant, which is a rather different approach than real reality, where we focus sight on objects located at different distances from us.

In the case of combination of devices based on the brain–computer technology such as: Emotiv EPOC+ NeuroHeadset with VR goggles, it is worth keeping in mind that artifacts may occur if both devices are placed close to each other. VR goggles operation causes small changes in operation of Emotiv EOPC+ NeuroHeadset, which in the end allows to control the character using the virtual reality. There are currently simulator prototypes using a combination of these devices. It is possible to move in these simulators by using generated internal events that to some extent can be seen as thoughts.

As indicated by the literature review, it should be possible in the future to use electromyography (EMG), i.e. diagnostics of the electrical activity of muscles and peripheral nerves, for determination of body behaviour in a given moment, and converting this information into movement of the avatar. Moreover, the proposed hybrid solution can be supplemented with EOG, i.e. electrooculography that recognizes data on resting potential around eyeballs in order to verify the current looking direction in the virtual reality that surrounds the person. There are also attempts to combine BCI-based devices with augmented reality (AR) [3], which has been presented in this chapter. It is worth noting that neurogaming is currently widely used in treating mental disorders such as attention deficit hyperactivity disorder (ADHD) [1], and post-traumatic stress disorder (PTSD). Increased interest in neurogaming in the world has resulted in organisation of the periodic conference held in San Francisco, USA, where the topics related to the ones mentioned above are discussed. Neurogaming, just as other practical applications of the brain–computer interface technology, raises ethical controversies. There are dilemmas concerning potential gaining/taking control over human mind by a machine or an individual. However, in the perspective of brain fitness, it becomes a promising tool, which was confirmed by the tests carried out [2].

References

1. Karimui, R.Y., Azadi, S., Keshavarzi, P.: The ADHD effect on the high-dimensional phase space trajectories of EEG signals. Chaos Solitons Fractals **121**, 39–49 (2019)
2. Paszkiel, S.: Control based on brain-computer interface technology for video-gaming with virtual reality techniques. J. Autom. Mob. Robot. Intell. Syst. (JAMRIS) **10**, 3–7 (2016). https://doi.org/10.14313/jamris_4-2016/26
3. Supan, P., Haller, M., Stuppacher, I.: Image based shadowing in real-time augmented reality. Int. J. Virtual Reality **5**(3), 1–10 (2006)
4. Wei-Yen, H.: Brain-computer interface connected to telemedicine and telecommunication in virtual reality applications. Telematics Inform. **34**(4), 224–238 (2017). https://doi.org/10.1016/j.tele.2016.01.003

Chapter 12
Computer Game in UNITY Environment for BCI Technology

Currently, the market of available technologies for creating computer games is very large and is still dynamically expanding. 1990s were a breakthrough in the history of computer games development when network games and 3D computer graphics were introduced into the games. 'TIE Fighter' space simulator from 1994 was the first representative of this technology. Presently, the Unity engine for creating computer games is one of the most popular programming environments, the first version of which, i.e. Unity 1.0.0, was created by: Dawid Helgason, Joachim Ante and Mikołaj Franciszek in Denmark and launched into the market on 6 June 2005. Not long ago new technologies, discussed in the previous chapters of this monograph, such as Augmented Reality (AR) and Virtual Reality (VR) entered the marked and are still dynamically being developed. The brain–computer interface technology is the latest technology which slowly makes its way into the field of games.

The Unity engine for games offers a lot of options which facilitate the creation of games or their assets. The basic construction elements in Unity are: game objects, components, variables. Game objects are any types of the content beginning with 'GameObject' and also including: light, visual effects, decorations, props, characters and also bigger objects such as the whole house or even a city etc. In order to create an object one must press the key create an object or it can be transferred from another program as a model. The game object in itself does not contain any action. To make the object useful and capable of performing actions, some properties must be added to it through components. Components define and control the behavior of the game objects to which they have been attached. The components can be compared to ingredients and the game object to a container in which the ingredients are stored. Some components are received automatically once we create or add a game object to the editor, like for example, component 'Transform'. Manipulating the components of such an component, in reality we manipulate the object itself. Variables are the properties that can be modified in the components. Like the components are elements in the objects (above-mentioned container), the variables are elements in the components. These properties may be edited using the window 'Inspector' and/or

© Springer Nature Switzerland AG 2020
S. Paszkiel, *Analysis and Classification of EEG Signals for Brain–Computer Interfaces*, Studies in Computational Intelligence 852,
https://doi.org/10.1007/978-3-030-30581-9_12

using scripts. Summing up these three basic construction elements made available by Unity, an object in a game is a constant element, a component tells an object what it is to do, and variables define how to do it.

In this game project for the purposes of neurogaming development, carried out within the research work of the Laboratory of Neuroinformatics and Decision-Making Systems of Opole University of Technology, named "Labirynt", Unity 2018.2 model was used, which was introduced onto the market at the beginning of July 2018. At the start of the project it was the latest available version of Unity program, therefore it was selected to be used. The Unity editor has many various functions and is very flexible. It is made to work under MS Windows and Mac OS X. It is used for animation, graphics, optimization, project management, sound, 2D and 3D physics, writing scripts in correlation with MS Visual Studio, etc. So many functionalities are possible owing to the flexible approach to creating this editor. Unity editor is actually a set of various windows of above-specified functionalities, operating within one application, i.e. Unity application. The most commonly used windows are the windows from the tab 'Window' → 'General': scene view, game view, hierarchy, inspector, project and console.

Scene view is the window in which a programmer creates a game. He can do it by drawing and dropping various objects from the window 'Project'. The scene window also has a set of tools controlling navigation, which are helpful in fast and efficient navigation in this window and they help in object positioning in the world of the game. Game view is the window in which a programmer can see what the game will look like in the end. He can check whether everything is working the way it should. Shortly, still unfinished product, game can be tested and keep track of what has to be improved and/or added. Hierarchy window shows all elements that are currently present in a given scene. Hierarchy is automatically updated once an object is added to the scene. Inspector window shows the components and variables of each object.

In inspector the components such as: scripts, collisions, sounds, object physics, colours, special effects etc. are added. Modifying the variables of these components it is possible to specify how the particular components are to function and thus, how a given object is to behave. Project widow is the window in which all elements of the project are collected. Each new element added to Unity will appear in the project window. These can be single prefabs, scripts, audio files as well as entire folders with a big amount of content. Console window shows all errors, warnings and other information generated by the Unity engine related to a given scene. A programmer may also display his own messages using functions Debug.Log, Debug.LogWarning and Debug.LogError.

The first reference to the computer game created based on the brain–computer interface and which was not only the 'brain training' comes from 2009 and is the MindGame. The MindGame uses P300 waves [1], which are the component of Event-Related Potential (ERP) connected with the process of decision making, for controlling a character on a three-dimentional board. The P300 waves are assigned to the particular movements of a character through linear selection and classification scheme. During network running the classification coefficient at the level of 0.65 was achieved, which ensures a gradual power of feedback reaching a player. Controlling the movements of the whole body is already known from Wii Nintendo, Sony PlayStation Move or Microsoft Kinect. Since this market is constantly developing dynamically, the prices of various motion sensors go down, which enables many companies a dynamic development of the games and devices using them and players to purchase the equipment and better game experience.

Hence, neurogaming is a relatively new area of the games using the brain—computer interface. Non-invasive methods of implementation of the brain–computer interface devices and reading the waves (EEG) are used to provide better and deeper impressions from the game. Owing to that, a potential user, player may interfere with the world of the game without using traditional peripheral devices such as a keyboard, mouse or a joystick. Zack Lynch (the founder of the Neurotechnology Industry Organization), at the conference 'NeuroGaming Conference and Expo' in San Francisco in May 2013, explained the term 'neurogaming' as a 'technology which all the time integrates the nervous system in the game, not only our brains'. These are technologies which monitor the pulse of a player, analyze facial expressions, register pupil dilation, hand gestures, body movement and changing emotional and cognitive states. All this is to provide a unique experience from the time spent in computer games.

To realize neurogaming within the 'Labirynt' game project Emotv EPOC+ NeuroHeadset by Emotiv Inc. was used, with which the Laboratory of Neuroinformatics and Decision-Making Systems of Opole University of Technology is equipped. The Emotiv EPOC+ NeuroHeadset interface is equipped with: 9-axle inertial motion sensors which track absolute change of position and orientation; Bluetooth Smart, which ensures a high quality wireless transmission owing to which a user has a full range of movement; an option to be configured by a user which enables the use of additional applications; a lithium battery of 12-h operating time. A few programs provided by Emotiv Inc. were used for correlation of the brain–computer interface system available with Emotiv EPOC+ NeuroHeadset with the 'Labirynt' game. EPOC Control Panel is one of them. It is the control panel which is to help a user in connecting the BCI Emotiv EPOC+ NeuroHeadset interface with a computer and configurating it. This panel makes it also possible to verify our calibration using in-built programs. The main window of the program is presented in the previous chapter. Another application used in practice was: Mouse Emulator Module of the EPOC Control Panel. In the module of the EPOC Control Panel, Mouse Emulator may be controlled with a cursor using a head movement. The operation of the module may be activated or disactivated at any moment. There is also an option of setting a preferred sensitivity of the cursor. Mouse Emulator was used in the 'Labirynt' game project as an application controlling the main camera which ensures the view of a player and the world of the game. It means that it is enough to move the head to admire and explore the world of the game around a player. This procedure was possible only due to the fact that moving the camera was programmed in such a way as to make its moving possible using the cursor of the mouse. Next, within the project carried out Emotiv Xavier Control Panel (Fig. 12.1) was used which ensures in a real time information pertaining to the quality of connection of Xavier and the interface sensors. It also provides the possibility of managing the player's own profile.

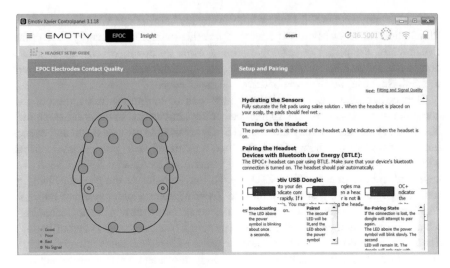

Fig. 12.1 Interface of the Emotiv Xavier Control Panel application

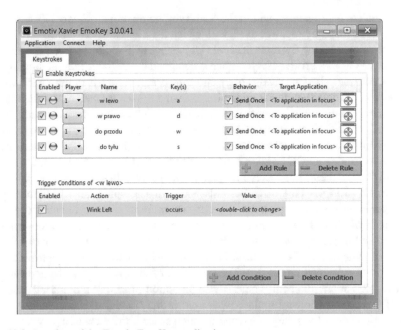

Fig. 12.2 Interface of the Emotiv EmoKey application

Emotiv Xavier Control Panel was used as a connector between the BCI interface and the EmoKey program in designing the 'Labirynt' game. This procedure ensured stability of connection and cooperation between the programs, the device and the game. Emotiv Xavier EmoKey (Fig. 12.2) is an application which can be combined with Emotiv Xavier Control Panel. It enables combination of the received results of event detection into any combination of the keys according to logical rules which are easy to define. EmoKey operates in the background and may operate any application which is currently in 'focus'.

Figure 12.2. shows the EmoKey application ready for use in the 'Labirynt' game. At first four rules were added, which consist of: enabled—enables or disables a given rule; player—identifies which interface is connected with a given rule; name—is a user-friendly name of the rule; key—combination of the keys performed while deploying a rule; behavior—controls whether a given combination of the keys is performed once or many times; target application—defines in which application a rule is to operate. Each rule corresponds to a given tack, action performed in the game (Figs. 12.3, 12.4 and 12.5).

Each rule must also have its condition or conditions of execution. In the rule named 'to the left' (Fig. 12.3) the condition for the event occurrence is blinking with the left eye, and in the rule called 'to the right' this condition is blinking with the right eye (Fig. 12.3), in the rule 'foreward' the condition is 'push' (Fig. 12.4), and in the rule 'backward' the condition is 'pull' (Fig. 12.5).

To create the 'Labirynt' game one external asset 'Cartoon Temple Building Kit Lite' was used. This asset was published on 17 February 2018 and was created by Anssi Kolmonen. Since the asset was made available in the Asset Store, one update was added on 9 March 2018. The asset is compatible with Unity 5.4.0 or later versions and contains 429 various prefabs such as: blocks, ground, gate, wall, etc. The entire asset was imported to the project and the prefabs available were used for the creation

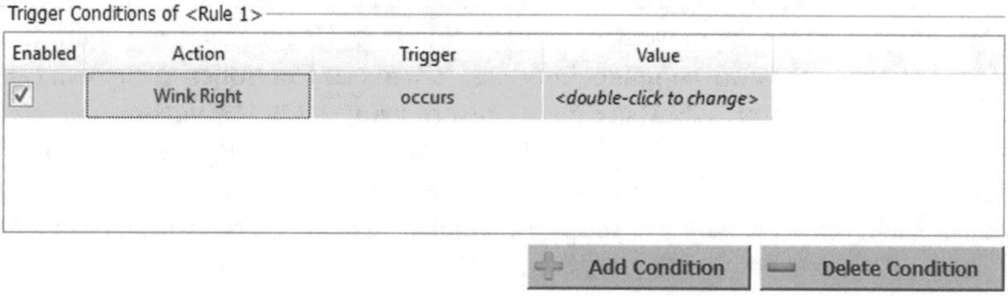

Fig. 12.3 Condition for the occurrence of the rule 'to the right"

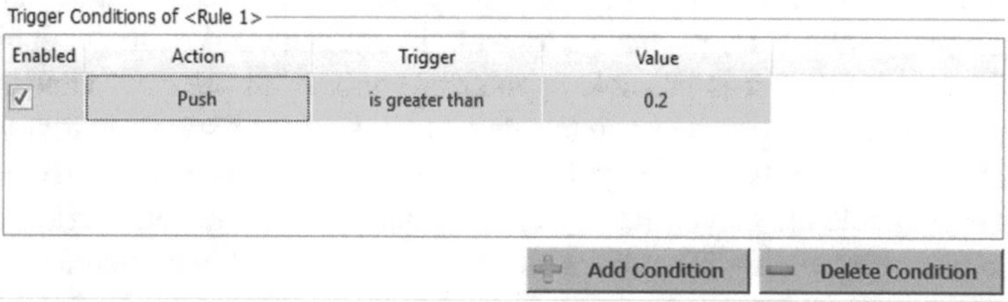

Fig. 12.4 Condition for the occurrence of the rule 'foreward'

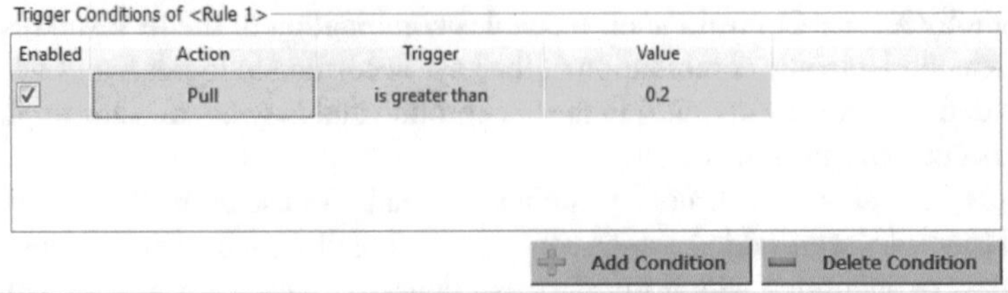

Trigger Conditions of <Rule 1>

Enabled	Action	Trigger	Value	
✓	Pull	is greater than	0.2	

+ Add Condition   — Delete Condition

Fig. 12.5 Condition for the occurrence of the rule 'backward'

of the 'Labirynt' world of the game. The first element added to the game and which began the creation of the game was the floor on which a player could move. Next, external walls, which create the borders of the world of the game, were added and later the internal walls, which as a whole create the labyrinth. The next element added were decorations in various places of the labyrinth, such as: plants, ruins, all kinds of uneven areas, etc. Decorations, i.e. decorative elements, are very important while creating games, not only computer games. Next, a ceiling was added, which gives the impression as if a player was underground and is to stimulate a player's imagination and impressions. Going through the labyrinth a player is looking for the exit from the trap he is in. Light was the last but one element added to the game. Light is a very demanding and time-consuming element of game creation as a programmer must take into account: distribution of light, its color and intensity. The final element of the game was adding the very player, i.e. the object with which a user will be playing. This object is an ordinary ball. A ball is a simple object and therefore it is easy to focus on it. Furthermore, moving a ball is easier because imagining a ball in movement is relatively easier than the movement of other objects and it enables easier controlling of the ball with the brain–computer interface. 'Character Controller' component and 'Movement' script were added to a player ensuring physics of the player in the world of the game and controlling the player. In order that the game may operate properly, 'colliders' were added to each object, which enable the player to get through the walls. 'Labirynt' uses only one script 'Movement' which is responsible for moving a player. It was written in MS Visual Studio 2017 by Microsoft in C# programming language because only this language is operated by Unity 2018.2.

Figure 12.6 shows variables which were used in 'Movement' script. The variables are divided into two sections, and then each section is also subdivided into two subsections. The first section is constituted by the variables which have impact on moving a player and here, we have a division into private and public variables, i.e. the variables in which it is possible to interfere only from the code view, and the variables in which it is possible to interfere from the side of Unity. The first variable is 'm_CharacterController' and it will be a component which was added to the player. Its task is to enable the player moving on 'colliders'. Next, 'moveDirection', which is a structure representing 3D vectors and points. It will enable orientation in space of three 3D vectors. Finally, there are three public varieties 'speed', 'jumpSpeed' and 'gravity'. These three variables, as shown by their names, refer to the speed of the player's moving, jump speed of the player and player's gravitation, i.e. how fast

```
 8    // player's moving variables
 9    private CharacterController m_CharacterController;
10    private Vector3 moveDirection = Vector3.zero;
11    public float speed = 5.0f;
12    public float jumpSpeed = 8.0f;
13    public float gravity = 20.0f;
14    // camera's moving variables
15    private float yaw = 0.0f;
16    private float pitch = 0.0f;
17    public Camera cam;
18    public float speedH = 2.0f;
19    public float speedV = 2.0f;
```

Fig. 12.6 Definition and initiation of variables

```
20
21    void Start()
22    {
23        m_CharacterController = GetComponent<CharacterController>();
24        gameObject.transform.position = new Vector3(-61, -7, -47);
25    }
26
```

Fig. 12.7 Start() function

he will be falling, respectively. The second section of variables contains the variables related to controlling the camera. Figure 12.6 shows two private variables: 'yaw' and 'pitch', which are responsible for the rotation about the horizontal and transverse axes of the player and the rotation about his vertical axis. The next variable is 'camera', to which the camera from the game was added and owing to which this camera could have been programmed. The last two variables are 'speedH' and 'speedV', which define with what speed the camera will be moving horizontally and vertically.

Figure 12.7. shows 'Start' function. 'Start' function is always performed at the moment of launching the script. In code line 23 (code lines to the left side of the picture) to variable 'm_CharacterController' was assigned 'CharacterController', i.e. the component which was earlier added to the player and which enables moving on 'colliders'. In code line 24, the initial start position of the player was assigned, which means that the player will begin the game in the same place after each new launching of the game.

Figure 12.8. shows Update() function. Update() function is performed all the time every frame until the script is disabled. From code lines 30 to 41, a conditional instruction 'if' was used so that the script might check whether the player is in the air or on the ground (m_CharacterController.isGrounded). When the script recognizes that the player is on the ground, the code lines which are inside the conditional instruction can be performed. And thus, in code line 33, a new 3D vector is assigned to 'moveDirection' variable, which enables the player's moving in

```
? Assembly-CSharp                                          ▾   ؟ٍ Movement
  28    ⌐     void Update()
  29    |     {
  30    ⌐         if (m_CharacterController.isGrounded)
  31    |         {
  32
  33                  moveDirection = new Vector3(Input.GetAxis("Horizontal"), 0.0f, Input.GetAxis("Vertical"));
  34                  moveDirection = transform.TransformDirection(moveDirection);
  35                  moveDirection = moveDirection * speed * Time.deltaTime;
  36
  37    ⌐             if (Input.GetButton("Jump"))
  38                  {
  39                      moveDirection.y = jumpSpeed;
  40                  }
  41             }
  42
  43             moveDirection.y = moveDirection.y - (gravity * Time.deltaTime);
  44
  45             m_CharacterController.Move(moveDirection * Time.deltaTime);
  46
  47             yaw += speedH * Input.GetAxis("Mouse X");
  48             pitch -= speedV * Input.GetAxis("Mouse Y");
  49
  50 ✎ ▌        transform.eulerAngles = new Vector3(Mathf.Clamp(pitch, -30.0f, 170.0f), yaw, 0.0f);
 8 %  ▾
```

Fig. 12.8 Update() function

3D world. 'Vector3' assumes three arguments: (Input.GetAxis("Horizontal"), 0.0f,
Input.GetAxis("Vertical")). The first argument returns the value of the virtual axis
called 'Horizontal', the second argument is set to zero and the third argument returns
the value of the virtual axis called 'Vertical'. 'Horizontal' and 'Vertical' input data
were set in the project in Unity environment and therefore the movent of the player
is possible in axis X and axis Z. When the player is in the air, i.e. he also moves in
axis Y, the player's moving is impossible. In code line 34, to the variable with 3D
orientation, the position of the object of the player is added. In line 35, the speed of
the object movement was added to this variable and everything was multiplied by
time. Then another conditional instruction takes place, which enables a jump when
the player is on the ground. It makes it impossible to make a double jump during the
game. In line 43, gravitation was added to the player's position, with which the player
will be falling during a jump from an object in the game. In order to execute these
dependencies on this particular object (player, ball), an element which will control
it is needed, and this element is a variable with component 'CharacterController'. It
is shown in code line 45. Then the behavior of the camera was programmed. In lines
47 and 48, the speed of rotation and in what axes they are to work were added to the
variables responsible for the camera rotation. Mouse X and Mouse Y are also input
data set in project settings in Unity environment and they are assigned to the cursor
movements in planes X and Y. It enabled controlling the camera with the movement
of the head by Mouse Emulator. In the last code line, rotation of the camera on axis
X and axis Y (axis Z is defined as 0 because we do not move the camera in this axis)
was added to the camera position. 'Mathf.Clamp(pitch, −30.0f, 170.0f)' function
limits the camera rotation so that the camera will not go under the floor.

Figure 12.9 shows a picture of the educational game in the form of a labyrinth
used as a basis of the frame for creating the structure of the labyrinth in the game
'Labirynt' and a screenshot of the game floor. The floor consists of a few blocks

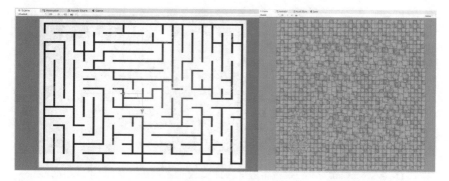

Fig. 12.9 A picure of the labirinth educational game and a screenshot of the game floor

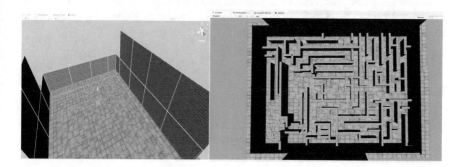

Fig. 12.10 A screenshot of external walls and a screenshot of internal walls

combined together and some deformations have been added, which can be seen in the figure shown below.

Figure 12.10. shows external walls in the game, which form the borders of the world. The player may not go beyond these walls and everything within the walls is the game world. Figure 12.10. also shows the structure of the labirinth, i.e. the internal walls, which form the whole shape of the labirinth. The shape of the labirinth is very similar to the shape of the labirinth in Fig. 12.9.

Figure 12.11 shows the game ceiling. It is constructed of the same big blocks arranged and rotationally selected so as to form one whole and imitate the collapse. White spots can be seen between the blocks. These white spots are crystals which will give light in the game.

Figure 12.11 shows distribution, color and light intensity in the game. Moreover, Fig. 12.11 shows what the light near the ruins looks like and the light given by the crystals on the ceiling, which are the main source of light in the 'Labirynt' game.

The main aim of the research works carried out and described in this chapter, which was achieved, was to create a computer game in Unity environment in correlation with BCI technology, The game was to give pleasure to the player drawn from exploration of the world with unusual possibilities of moving in this world. This

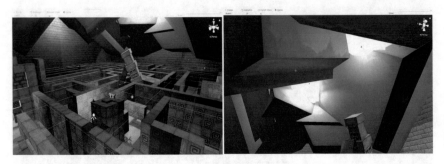

Fig. 12.11   A screenshot of the labirinth light and a screenshot of the crystal light on the ceiling

objective has been achieved. This game has also another advantage resulting from the control system used as focusing is required for moving. Hence this game has an additional pro-development effect in thinking processes.

Reference

1. Jotheeswaran, J., Singh, A., Pippal, S.: Hybrid video surveillance systems using P300 based computational cognitive threat signature library. Procedia Comput. Sci. **145**, 512–519 (2018). https://doi.org/10.1016/j.procs.2018.11.115

Chapter 13
Using BCI in IoT Implementation

In this chapter, based on the development work carried out, the solutions making it possible to combine a few IoT devices and control them using BCI technology are presented. It consists in sharing a platform with a user within which he will be able to connect all the devices possessed with one common data exchange channel [1]. These data in the implementation presented will be used to change the state of the device or a group of devices as a reaction to the event taking place in another device. The system developed will show a high level of abstraction and universality so that it may be used with a wide range of quickly developing devices without any significant interference with the most important elements of the system itself. Furthermore, during the system implementation, special attention will be paid to the use of the state-of-the-art technologies and trends in creating software in such a way as to avoid the system becoming outdated even in the time of quickly developing technologies connected with producing software. Within the research carried out, a frame of the system which will allow for operation of as many devices as possible with exemplary modules operating within this system was made. These models will allow for visualization of the application possibilities of the platform created based on the example of controlling a home audio device and electric sockets using a mobile Android device. The application operating the device and being a part of the system will have, additionally, to possibility to connect and use for controlling brain–computer interfaces created based on the device reading the EEG signal (electroencephalography) of the brain and subsequently processing it into relevant events.

Apart from the solutions strictly oriented on controlling IoT devices, attempts to integrate IoT devices with BCI technology have already been made, the aim of which is, similarly as of the system designed, to make it possible to control the surrounding devices using only the signals coming directly from the brain [2]. The company Blackrock Microsystems Inc. working on the development of BrainGate technology has made the biggest progress in creating such a system. It includes creating implants reading data coming from the brain and the equipment and software processing these data into computer commands. Authors foresee its integration with the devices such

© Springer Nature Switzerland AG 2020
S. Paszkiel, *Analysis and Classification of EEG Signals for Brain–Computer Interfaces*, Studies in Computational Intelligence 852,
https://doi.org/10.1007/978-3-030-30581-9_13

as a pneumatic arm helping paralyzed people to perform their everyday duties using thoughts. However, the system is not available for everybody, because the target group of the operated devices focuses around the professional equipment for the disabled and not the household IoT appliances. In this approach, it is the system that is in the centre and the equipment, to some degree, is adjusted to cooperate with it. A significant disadvantage of this solution is an invasive method of reading the EEG signal with the electrodes implanted directly into the cerebral cortex. It makes the collection of a significantly more accurate signal possible and therefore much more precise device control, however, at the price of interference with the user's structure. This solution also causes other additional problems such as an implant ulceration progressing with time. The use of the invasive method also creates new possibilities with a very precise and safe controlling but it also limits the possibility of application for commercial purposes. Thus currently, there is a gap in the market between the solutions combining BCI interfaces with IoT devices for a commercial user.

The Internet of Things has already an almost 20-year history. The IoT development has a strong impact on people's lives and the surrounding world but also carries a lot of threats which, maintaining the same pace of development, may become a serious problem which the users of this technology will have to manage. It is hard to tell when the technology of the Internet of Things has its beginning. Most sources provide that the event which gave the beginning of the Internet of Things was RFID (Radio Frequency Identification) developed by Kevin Ashton in 1999. A simple system, which was to replace a well-known barcode, made it possible to connect a few sensors and actuators together and gave the foundation for the Internet of Things. Also, it was Kevin Ashton who introduced this term in his works.

The concept of the Internet of Things got foothold and manufactures began integration of their devices with the Internet network. They did not limit themselves only to refrigerators, but a whole range of devices became available from the network level. The imminent problem was noticed quickly and it was an insufficient number of public IP addresses. IP version 4 addressing used until now assumes the possibility of creating no fewer than 300 million unique addresses, which were running short when each ordinary device had to be assigned a public IP address. This problem was usually solved by creating local networks of internal addressing, but it caused lack of access or difficult access to a device outside this network. The breakthrough in the development of IoT was the moment when the addresses IP version 4 began to be replaced with IP version 6 addressing. The new addressing provides for a big number of available addresses: 1028, which may be assigned to refrigerators, cameras or simple sensors without any concerns. A revolutionary step, bringing the Internet of Things to the current state, was transferring many solutions from this area to the cloud. This allowed for a considerable simplification of these devices which, since then did not have to have their own data repositories or a local server dedicated to their operation. Instead, they can exchange data and commands via always accessible, highly scaled cloud. The use of this cloud, additionally, allows filtering and combining data coming from many devices for better understanding of the situations taking place by the systems and IoT devices. Clouds can also be equipped with the solution from the artificial intelligence field, which, based on the data received, will help in better reacting to the existing situations, and thus better integration with the user and adjustment to his needs and requirements.

The system connecting a considerable number of devices, which is by definition a distributed system, must consist of many elements, which are external elements from the point of view of the system. Because the system should be possibly universal and operate a wide spectrum of external devices, these elements may be replaced with other elements fulfilling similar functions. In this chapter, elements of the system grouped according to the function they fulfil are described. Additionally, at least one alternative to a given element or device for each of the device mentioned is presented.

For an application operating only on a limited functionality provided by a device, the use of Emotiv SDK offering high-level data coming from the device, without the need of a manual analysis of the user's EEG signal is much more convenient. Emotiv SDK in Com- munity Edition offering support for Android platform was used for interaction with the designed system. The version adequate for this platform offers a set of the Java language interfaces which allow, among others, for switching into the pattern learning mode and back. The device, when switched to learning mode, is able to remember a current brain activity as a pattern and assign it to a given, formerly selected, command. The learning algorithm of a new command is shown in a form of a sequence diagram in Fig. 13.1.

Teaching a command already learnt can be repeated at any moment overwriting a formerly learned pattern. Once a device is taught at least one command, it is possible to begin listening to the commands taking place. They are transferred by Emotiv SDK as events marked with the name of the command and additionally transferring strength of the detected event. In practice, the source of the events is one central function which, when polled in definite time intervals, provides information on the events taking place. A delay with which a system is notified on interaction of a user with a device depends on polling rate of this function. Actual delay between the performance of the action known to the system by a user and information about it reaching the system consists of: delay connected with sampling Emotiv EPOC+ NeuroHeadset, delay in Emotiv EPOC+ NeuroHeadset operation; delay in transmission of information via Bluetooth channel and operating this channel by both devices involved; delays caused by data processing by Emotiv SDK; delays resulting from time intervals between consecutive polling of Emotiv SDK function by the system.

Therefore, in terms of operation of the suggested system created within the development works on BCI technology it is important to balance correctly the function polling rate. On one hand it should be as high as possible to be able to react fast enough to the user's commands, but on the other hand, law enough not to slow down significantly the operation of a client application and not to reduce its responsiveness to the events taking place via different channels. A MindWave Mobile headset by NeuroSky might be an alternative for the device used. Like Emotiv Inc., it offers the reading of the brain EEG signal by the sensors placed in the appliance. Also in this case the manufacturer provides program tools in the form of Android Developer Tools, which has the functions as the library of its competitor. In order to add NeuroSky MindWave Mobile operation to the created system, an additional module of

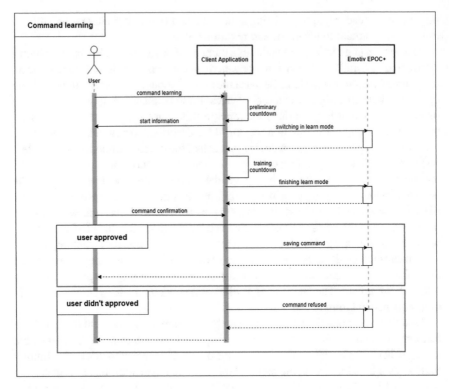

Fig. 13.1 Diagram of a learning sequence of a new command by a BCI device

the client application must be created to operate this headset and registered as an event deliverer.

A very significant step in software development was moving from traditional monolithic applications to distributed applications consisting of many elements. They can be represented by microservices, various processes on physical machines or even separate computers. The methodology of creating distributed systems brings many advantages: from a considerably better distribution of responsibility, through higher scalability to the issues of increased security. Such approach was also chosen in the case of the system designed, represented by executing devices, joined into a network constituting a system. The executing device is a computer designated for interaction with one or more IoT devices. It is necessary, especially for the devices which do not have the Internet connection and must be controlled directly or for the devices, which for security reasons or accessibility should be controlled from a local network and not from the central server. Electric sockets, which are controlled by transmitters, are representatives of the first group of appliances. A cable is routed from the executing device dedicated to such sockets which transfers control signals. A TV set with Smart TV functions may offer the possibility of interaction via the Internet, but the port on which communication with it takes place may be unavailable from beyond the local network. In both cases the solution is applying an executing device operating the sockets, television or both at the same time.

Owing to the launch environment and programing language, a mobile Raspberry PI 2 computer works best as an executing device, on which any operating system supporting NodeJS environment is launched. In the case of an exemplary environment, Raspberry PI 2 system was selected for the system designed. The executing device used has small dimensions, it may control voltage status of a universal port, does not consume too much power even when used continuously and is quite efficient.

As an alternative for the computer used also other versions of the popular Raspberry PI 2 computer with any operating system supporting NodeJS platform might be used, A miniature Intel Compute Stick computer operating under Microsoft Windows is a good choice as well. A PC (Personal Computer) class computer may be an executing device, which using in-built interfaces (USB, LPT serial port, COM port) or via the Internet may communicate with IoT devices. In order to use another solution instead of Raspberry PI 2 computer, a system instance should be launched on it, an encrypted connection with the central device must be ensured and a trusted public key of the device should be registered. Summing up, each executing device must meet the following conditions: it is in direct communication with the central device; system application instance of the system designed is launched; it is correctly registered in connected central devices and is on the list of admitted devices; it is responsible for interaction with one or a few external devices; it can operate at least one type of events or is the source of events; it identifies with the public key and conducts encrypted communication with other devices of the system.

A distributed architecture of the system causes a few problems. The main problem is lack of knowledge of the client application of the structure of the system and its particular elements. Central devices were applied in the system as a solution of this problem. They are computers fulfilling the function of a connector between executing and client devices on which the system application instance is launched. Central devices should be efficient as it is through them that the traffic from client applications and some part of their traffic connected with data flow between the executing devices pass. Furthermore, the central device address should be fixed and publically accessible as each of the client applications must have information on the address of at least one of the central devices.

A dedicated server may be an alternative to the shared server used for the users demanding higher performance from the system offered. Such a server will certainly show higher performance and therefore shorter latency. It is a server to which a user has full access, it can assume the role of a database server, which will result in a significant shortening of access time to the database. For the users expecting still higher security of their data, a possibility was provided for hosting the central part of system directly from the user's infrastructure. A computer operating NodeJS platform is required to be able to use this solution. Moreover, such a computer must have a fixed IP address so that it may be configured in the client application. If a user requests that the system be accessible also from outside the local network, the computer must also have an external address and the port on which the application listens must be accessible from outside the local network.

Actions which finally, via the system, will control the end devices may be evoked in a few ways. The devices the task of which is to evoke control events are called client devices. Evoking events may take place on them as a result of interaction with a user, a result of external device operation, such as already mentioned Emotiv EPOC+ NeuroHeadset or automatically. In the system discussed, client applications for two kinds of devices were implemented: for smart phones or tablets operating under Android 4.4 or newer version and devices equipped with the Internet browser meeting the requirements of Angular 4 framework. All of the above-mentioned components are used either for sending commands to the system or processing these commands and managing the system. However, the system remains useless without end devices which will be controlled by the system. It should be a group of devices whose manufactures provided for the possibility of controlling via a network. Therefore it can be concluded that connecting such devices to the system is useless as many of them are provided with a dedicated control panel which is most often an internet application hosted directly on the device itself. However, connecting such a device provides the possibility of unification of these panels and combining them into one common panel accessible under one address. Integration with the system also provides a key functionality of reacting to the change of the status of other devices.

There exist many ways of providing access to the control interface, but as of now, REST API is the most common one. This way of the external interface implementation is the easiest in operation in terms of the system. In the case of such approach, the only action required is sending a HTTP request of a relevant method to a relevant URL address. Sending HTTP requests is operated by default by the NodeJS platform used. Thus the information what action is to be performed in a target devices is a resultant of three request parameters: address, optional data and a method, where the method in most cases of implementations is one of the following: GET if the aim is to receive data; POST if the aim is to sent data for request; PUT to exchange the existing data for the data sent in the request; DELETE for data removal.

Apart from the devices making API available, whose integration is limited to programming solutions, also traditional devices are also often seen in households. Here, we can mention traditional bulbs, electric blinds, heating or air conditioning. There are many reasons why a potential user might also want to control the devices from this group using the system designed. An example of such a scenario may be a wish to adjust lighting to the time of the day or the possibility of central controlling the level of lowering the blinds in the whole house.

The implementation of the system in the current scope requires, apart from the hardware solutions, also many programming solutions and tools that are not parts of the system. Before starting work on the application, the requirements which the application should meet were established. The system is to meet the following requirements: it should provide for control using a mobile device; it should provide for integration with a device offering the BCI interfaces; it should offer integration with basic IoT devices; it should allow the use only by authenticated users; it should make it possible to add new system users and managing them; it should offer the possibility of extending the range of the operated devices in the future; it should ensure security of the data flow; it should be scalable in an easy way; it should be created based on

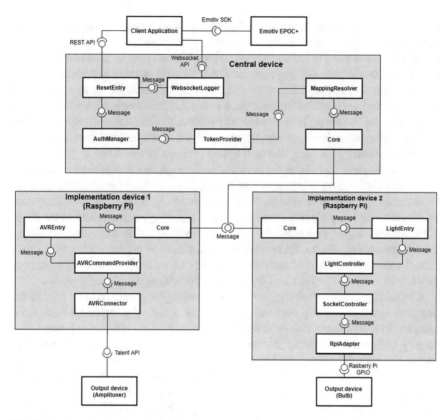

Fig. 13.2 Diagram of components of the system created

state-of-the-art technologies and patterns for program developing. The stage of the system practical implementation was preceded by the phase of the system architecture design. Special attention was devoted to this phase due to the requirement of universality and scalability. A diagram of components, which is the outcome of this phase and at the same time documentation of the way of the system construction is shown in Fig. 13.2.

As shown in Fig. 13.2, the system is based on the architecture of microservices the main aim of which is breaking up a monolithic application fulfilling many functionalities into separate, possibly small services satisfying the single responsibility principle. The services should be independent of each other, which practically means that one service may not create directly the instance of another service. Such an approach has numerous advantages such as: easier creation of new functionalities through division of responsibilities; possibility of cooperation of many people or teams without a need of frequent synchronization of progress made; independence from one programming language and technology and possibility of combining various technologies; possibility of dividing the system into various devices or containers; possibility of combining software of various versions and easier update of single services. Apart from numerous advantages, treating software as a set of separate services also creates new problems which were not known while creating monolithic applications. The essential one is a need to inform a client with which service he is to communicate. A few solutions to this problem were invented but none of them is perfect.

Creating a component fulfilling the function of a register is the solution used most often. This register contains information on all other components. This information may find its way to the register through registration, by each launched component or through finding all active components by the register. However, it is not a solution easy in implementation or scaling due to frequent changes of the status of IoT devices. It enforces passing the whole network traffic through one component which, as a result, quickly becomes a bottleneck of the application. Also the register data must be stored in an efficient way so as not to slow down significantly the whole system by this additional step. Another possible way of adjusting microservices to the needs of the internet application is the use of fixed addresses for the components making the architecture similar to the monolithic application. However, such an approach excludes many advantages of microservices: they are difficult in changing a version, they are no longer so flexible and do not allow for adding new functionalities.

A hybrid approach, combining the two methodologies presented were used in the designed system. The register was given up for a central component with fixed address. This component does not return information on other components but directly redirects a request to relevant services. The components still have to register in the central component, but until establishing connection ensuring bilateral communication channel. As long as this channel remains open it is deemed that the service on the other side is accessible. This channel is used for sending a request and a response to this request and also for exchanging information on the requests which can be serviced directly by the service. Giving up the central register the possibility of the central service duplication was achieved, which makes the application scalable widthwise.

Practical implementation of the architecture presented has been based on the services launched on the Node platform. A service, or a few services, are launched on it in a container of applications delivered as a part of the system framework. It fulfills the following functions: it automatically allows for dependency injection and implementation of the inversion of control pattern; provides HTTP interface support providing services with the possibility of sharing API available from client devices; provides WebSocket support allowing the creation of a bilateral communication channel with client applications or directly with other services; connects with the central component using a client's WebSocket protocol; automatically attempts to resume interrupted connection with the central component; connects with client components using the WebSocket server; provides an object of the class responsible for unattended communication with the rest of the system; provides connection with MongoDB; provides an object of the class responsible for cryptography: encrypting, decrypting, signing and data verification; provides the possibility for injecting a finished configuration object to each of the modules; provides classes and interfaces normalizing the implementation of modules.

Thus a container in the system designed is a facilitation for service creators which allows to avoid multiple code duplication and normalize using shared resources, in which it is easy to create name conflicts. Dependency injection is an essential functionality offered by the container. It is so important for the whole system because it offers the possibility to inject classes responsible for protocol support, database management or cryptography. Dependency injection is not supported by default by Node so in order to be able to use it was necessary to use a library providing this functionality. Dependency injection underpins the IoC (Inversion of Control) pattern assuming giving up the creation of class objects directly in other objects in favour of providing them from the central place, which allows a loose combination of both classes and removal of the dependence of the class creating an object on the class of the object created. Instead, the objects are created in one point, in which it is possible to decide on the lifespan of each object, implementation used to create an object and even multiple reuse of the same objects. *createContainer.ts* file fulfils the role of the central point in the created application. It is configured by default in such a way as to allow injecting the objects of all basic classes of the container and the ones which are in the folder *modules*, up to the second level of nesting. The objects are created as singletons, which means that they will be created once, during their first use, and further calls will be using this instance. To unify the way of access to the objects, camelCase notation was used. To use dependency injection in the model created it is enough to add as an argument, in the constructor of one of the classes, an object which has a field corresponding to the name of one of the accessible objects. HTTP protocol operation is the task often performed by the applications operating on the Node platform. In order to create API for communication with client applications, the tools embedded in the platform may be used or it is possible to use one of the libraries facilitating this task. In the designed system, Express library was used, owing to which the creation of basic API comes down to port configuration and then definition of the resources available. Express library, after initial configuration, which includes setting of the port listened and connecting a library parsing cookies and request bodies and returns the object based on which the resources can be defined. This object, owing to dependency injection, is accessible from the level of each class which, by using it, can define the module resources.

HTTP protocol sharing by many modules causes the possibility of occurrence the name conflict within the resources. To prevent it, but also to precede all resources with one, common for the whole system prefix, a convention was adopted that all modules should precede the declared resources with the name of the same module.

From the point of view of client applications, WebSocket protocol is mainly an addition to HTTP protocol allowing communication in the opposite direction than HTTP, i.e. from the server to the client application. Sending data to the connected client application is carried out by injecting a WebSocketServer class object and calling the broadcast function receiving the data sent as an argument.

Current implementation only provides for sending data to all clients connected. The possibilities of the protocol and the supporting library are much bigger and make it possible to send data to a particular client or to upload a list of active clients. The problem encountered during the implementation of this protocol operation turned out to be the configuration of the server used. It has mechanisms checking if a connection is active and in the case of lack of activity it ends this connection with an error. It is not an expected behavior for the data exchange channel, which by definition should remain open during the whole session of the client to allow the receipt of possible communications coming in the direction from the server to the client. Such behavior made it necessary to artificially maintain the activity of the WebSocket channel by regular sending *ping* communications. These communications sent by the server do not require a client's response but allow to maintain his direct connection with the server. The WebSocketServer class makes the broadcast method available, which makes it possible to broadcast data to all active clients.

Communication within a system is vital in many respects. It has a significant impact on the performance of the whole system and is one of the main candidates for the target of the attacks on the system and it must be adequately universal to manage the tasks set by equally universal, modular and distributed system. Summing up, the requirements pertaining to this part of the system are as follows: high speed of operation; possibility to be separated as a reusable module; easy operation, possibly low code input required for operation; operation of synchronous and asynchronous requests; possibility to operate both in a local environment and remotedly via network; management of various types of data; possibility to modify server topography in a real time.

The communication carried out in the local environment, inside one container, is used very often in this type of systems. There exist many ready-made solutions: from EJB technology for Java Enterprise Edition, which is characterized by its operational speed and the possibility to type call arguments owing to the possibility to use the repository storing a set of key-value pairs in the operational memory and also internal mechanisms of communication between the containers operated by Docker. The application of an own container in the system, which provides the possibility to launch independent modules within one process, allows to apply an additional, most simple method—a direct function call. Obviously, direct querying of the service functions by another service definitely clash with the requirement of modularity imposed on the system as well as with the basic principle of the architecture of microservices and IoC pattern. This problem was solved by the application of an intermediary class. This way a module which wants to call the function of another module does not do it directly but only packs in a message describing its request and further delegates this message. The central component, based on the internal list of the known components, processes this request. This solution only insignificantly increases the request service time by finding the function which can operate it and it allows the separation of communicating modules.

Registration of modules, even though it seems to be an easy task, is connected with a considerable amount of additional logic. Both the possibility of real-time change of registration plate and registration of the modules available remotedly. Also the facility of this solution should be maintained so that the necessity of registration may not become disruptive and require major changes in the architecture of the modules. Therefore *pattern matching* in the form sufficient for finding a method that could handle a request was applied. In order to implement a message pattern, the requests that represent them were provided with three items of information: package name which should identify a module, and action parameters. The modules which can handle requests are also registered with providing these parameters and leaving a parameter blank is understood as the ability to handle requests of any value of this parameter.

Thus, the solutions used come down to packing requests in messages and registration of the functions handling messages of a given type. To facilitate the creation of messages, the library supporting internal communication makes the function returning the framework of such a message available. This framework is to be supplemented with data, relevant parameters, name of the packet and action, and then request its service.

Remote requests from WebSocket are supported by the system in a similar way as local requests. The type and format of the message, which is sent between the application instances was established, is treated as a remote possibility of module registration. The application in such a message, apart from a special header which identifies it, sends its own local list of standbys, which a recipient saves with information on the connection between the devices and the key identifying a device. Additionally, this way of registration allows for a fluent manipulation of the registration list through registration and deregistration of single modules even during the system operation. The central component, basing on the current registration list, can adjust its topology to the existing situation.

For remote applications, the most common solution is REST API sharing and remote function calling via HTTP. However, it is not the fastest solution due to additional data that must be attached to each request and the time necessary to initiate and close HTTP connections. In this respect WebSocket, which is a more low-level protocol than HTTP, is much faster, especially while sending a big number of requests.

The use of WebSocket, even though it increases the system performance, causes problems connected with another issue, i.e. synchronization of outgoing requests. WebSocket is an asynchronous protocol, i.e. the protocol which only sends information not waiting for the response code of the target application. The necessity to receive feedback requires additional support from the level of the application itself. In order to achieve it, messages are assigned unique identifying numbers and then a response of the same number is awaited. When such a message is received, the function calling a request is notified about this fact and the data from feedback are transferred. This solution makes it possible to send requests to any modules without the need of information on the address or even the module location. Therefore, for the requesting module it does not matter whether the target module is within the same container or even on another device supported by another network.

Like many other applications, also the designed system needs access to a persistent data repository. The main applications of databases in the system are: storing users' accounts, recording and reading server configurations; storing information on actions evoked by the occurrence of particular events; recording and reading a list of events and a list of actions; providing modules with the required data and saving them; support of the module sharing global data.

The data for the system might be stored in any kind of a database. However, non-relational databases such as MongoDB, CouchDB or Cassandra are supported exceptionally well by Node. This is an alternative for traditional relational databases, which is doing exceptionally well dealing with a big number of unstructured data. Instead of traditional tables connected with one another with rigid relationships established in the initial phase of database designing, the databases contain sets of documents called collections. The documents in non-relational databases can store all objects together with their fields, including even those representing the tables of other objects. The structure of the data stored can be changed considerably without the need to rebuild the whole database and lose the existing records. Node has a library to support Mango—one of the most common non-relational databases, which makes it possible to reduce the interaction of a programmer with the configuration of a database to minimum. In order to save data using the Mongoose library it is enough to provide the library with the address of a database and create the models corresponding to the templates of the element recorded in the collection. Using the function of JavaScript supported by Mongoose it is possible to start recording and reading data from the base. It makes it possible to start working instantly on a new collection without the need to create tables, relations and establish rigid data structures.

The integration of the data repository with a selected architecture of microservices turned out to be a challenge. The services, which are separate by definition, to be able to be launched on physically separate computers, sometimes may require access to the same data. Obviously, an ideal separateness of the modules may be assumed, also including separateness of the data they use, but in the designed system the necessity to access some configuration data and some common modules excludes such a solution. Therefore the possibilities to ensure access to the base in such a distributed environment are as follows: replication of the database for each container connected with the need of continuous synchronization of the data contained in them and excessive use of storage memory; creation of a module accessible with a mechanism built in the system and providing interface for data reading and manipulation in the common base.

The former solution, even though it seems more efficient, generates a big network traffic required for data synchronization. Also the issue of synchronization itself is such an advanced issue that in the case of the smallest implementation error it may lead to data jitter and thus a faulty operation of the system. The data used for synchronization would have to be additionally coded to disable a possible attack of the man in the middle type. The latter solution, even though it adds at each database request the necessity of additional polling of an external module, does not generate such a heave network traffic as in the first solution. Analyzing the effects of using each of these solutions, it was decided to use in the application a hybrid solution. The modules, which will be launched only on one device and which will be usually classified as the actuating device, may have their own data repositories launched directly on the same device. This will shorten significantly access time to the database and thus accelerate application running by the time of the additional request to another device. For the data which may be used by many modules, an additional module was provided, which acts an intermediary between the database and the application. This module, after recoding, reading, deleting or modifying the data from the request, responds to the requests coming from other modules.

For the purposes of a widely understood system security, two modules handling the issues related to cryptography: *CryptographyHelper* and *AuthTokenProvider* were designed. The former is to store the private and public keys of the application and code the data using them. The keys, which are the files generated with *ssh-keygen* command are packed in the NodeRSA library objects, which are easier to use. After the implementation of the keys, the library provides the functions for encrypting, decrypting, signing and verifying the signature of any text data. The module handling cryptography owing to the application of the IoC pattern is available as the object injected into the constructor of other modules, which makes it very easy to use. This ease of use was one of the key requirements demanded from this module as it is owing to this easiness that many data processed in the system are sent in the decrypted form.

AuthToken-Provider is another cryptographic module from this group responsible for managing the user's tokens. A token in the system is understood as a unique series of signs assigned to one of the system users and authorizing him to use this system. This token is generated by the above-mentioned module while logging in and then it is verified at each request of this user. It was decided to use JWT (JSON Web Token) tokens in this module. The basic feature distinguishing JWT tokens from ordinary random series of signs is their possibility of transferring the data which may be read without knowing the key. The module handling tokens, unlike the cryptography module, is not available through injecting into the constructor, but as a model available after sending a relevant type message with a request. Such a solution was imposed by the need to use the tokens generated with the same key within the entire application. By making the token generation available as a module, it is also available for other modules, local or launched within other containers, on other devices.

The system, mainly due to its modularity, universality and the technologies used, is characteristic of a significant complexity of the operations performed. Beginning with the application on a client device, through the central device and finally actuators, a control command covers a long distance before it finally reaches the IoT device. The issue of security is vital for the system operation. The system offered has access to numerous household appliances of the user. Very often, these appliances are not well secured and may easily become a victim of attacks. The essential issue related to the security of each application is the way in which it protects access by unauthorized persons. Additionally, the mechanism of authorization, i.e. recognizing if a user is the person he claims to be as well as authorization, i.e. verifying if this user is authorized to perform the action can be indicated here. The authorization in the designed system is based on using an authorization token. This way of authorization implementation is currently a standard in various types of applications as it is easy to implement and at the same time it avoids sending a user's password and it is possible to use it in distributed applications and the applications using microservices. In the designed system the role of the token is fulfilled by the JWT token generated by the *AuthTokenProvider* module. Apart for ensuring protection of the access to the system by unauthenticated users, an important issue showing the security of a system or it susceptibility to attacks is the way of storing the data authenticating a user in the data repository. Presently, *SHA-256* algorithms with additional salts or the *bcrypt* function, which was used in the discussed system, are recommended methods for password saving. The mechanisms of authorization were abandoned in the designed system as all the users were assigned identical authorizations pertaining to all parts of the system. However, if there should appear the necessity of dividing the users into groups of various authorizations, it would be an easy task to carry out, only through extending a user's model by additional roles and then verifying if a given role is possessed, if necessary.

Apart from an appropriate and secure way of user authentication, a secure system must be resistent to attacks aimed at the system infrastructure itself. The places in which data are transferred between devices are especially vulnerable to this kind of attacks. These places may be tapped by potential attackers leading to data leakage or, even worse, modified and thus make it possible to carry out the man in the middle attack. The attacks of this kind are especially dangerous as most often applications treat them as trusted data coming from other parts of the system and they neither validate these data nor verify them in terms of potential attacks. The attacker, who has access to the data transferred and the possibility to modify them and knowing the system implementation or trying to guess it is able to carry out a fatal RCE (Remote Code Execution) attack, and then nothing can prevent a complete system takeover, access to the data of its users or even going beyond the system framework and attacking entire server and devices hosting parts of the system.

A considerable increase of security of the transferred data may be achieved by using the Encrypted HTTPS protocol, but in order to increase security additionally and have a complete control over the place and the way of data decryption, they were secured additionally for the time of their transfer between the devices using asymmetric cryptography. Each container of applications has its unique set of keys: private and public and knows the public keys of the containers with which it is directly connected. Public keys of the central device by which the actuators recognize the configured central devices are in the configuration file of each device. The keys of the actuators known to the central devices are configured by the internet interface of the central device and recorded in the database. In this way it is possible to freely configure the network topology by changing the key lists of the connected devices. Apart from difference in configuration of the list of known public keys, other actions connected with cryptography are performed in the same way, both by the central devices and the actuators. The data leaving a device are encrypted using the public key of the target device and decrypted only in the target device using the private key. This way the data, during the whole way they cover between the devices, remain encrypted and data takeover by third party will not lead either to the leak of information or the possibility of their modification.

Thus the cryptography of the public key was fully applied in the designed system because each of the keys fulfils the following two roles: the private key is used for data decrypting and signing. Both roles complement each other securing all communication channels between the system components. The communication secured in this way may be deemed so secure that the participating data are treated in the system as trusted data.

The implementation of the system presented in this chapter was subjected to verification by carrying out a series of both performance and functional tests. The performance tests make it possible to verify how many users can use the system maintaining its smooth operation and adequately low reaction time. In order to obtain similar conditions to real ones, the central device was polled simultaneously from the computer from the home local network according to the scenarios presented below. All load tests were carried out using the loadtest application on the computer connected to 100 Mb/s Internet network. The first scenario carried out is to check only the access time to the central device. After receiving any request it returns immediately the HTTP code proving a successful performance of the test (Table 13.1; Fig. 13.3).

It can be read from the test results for this scenario that the system reaction time in this configuration remains stable even with the load that is unreal with so many simultaneous users. Real delays of 250 ms should not be noticeable for the application user. To decrease delay time in the above test, the performance of the server supporting the system or the internet bandwidth of the client's device should be increased.

Another test scenario carried out corresponds to a typical everyday use of the application. The test results, in the context of the results of the first scenario, allow

Table 13.1 Test results of the central device polling

Queries per second	Number of queries	Average response time (ms)
5	50	256.9
5	100	254.1
5	250	254.5
5	1000	256.1
10	50	254.5
10	100	252.5
10	250	254.7
10	1000	254.7
20	100	260.9
20	200	258.3
20	500	258.7
20	1000	255.0
50	250	255.5
50	500	263.5
50	750	260.2
50	1000	255.6

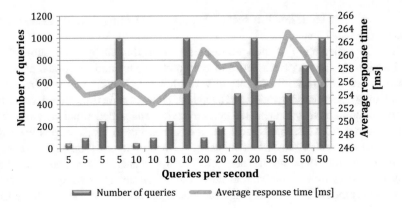

Fig. 13.3 A diagram of response time of the central device polling

Table 13.2 Test results of sending a request to the actuator

Queries per second	Number of queries	Average response time (ms)
5	50	259.2
5	100	256.7
5	250	254.2
5	1000	256.6
10	50	255.8
10	100	255.4
10	250	254.7
10	1000	256.6
20	100	260.9
20	200	258.3
20	500	258.7
20	1000	255.0
50	250	255.5
50	500	263.5
50	750	260.2
50	1000	255.6

for the measurement of the operation time of the system itself and therefore the quality of its implementation.

It can be concluded from the results presented (Table 13.2) that the delays generated by the system are negligibly small at such loads (Fig. 13.4).

Summing up, it can be stated that at the time of quickly increasing popularity of the IoT devices and all-present mobile devices, a demand for a system connecting equipment of various specification designed by different manufacturers is very big.

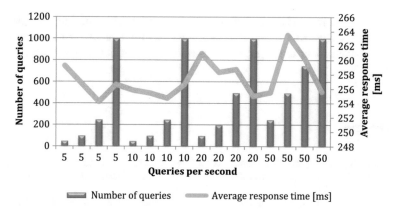

Fig. 13.4 A diagram of response time of a task delegation to the actuator

This division describes the implementation of the system of this kind and is an attempt to satisfy this new need. It shows that in this type of applications topicality of the technologies and solutions used, which finally decides how long a system can stay on the market before it becomes totally outdated, is a significant parameter.

The system designed is not a complete solution possible to be applied in this form on a large scale but demonstrates only its possibilities as a framework for creating final systems. First of all, it features a big universality which offers possibilities and indicates directions of its further development. The availability of modules is the main area in which the system may be developed. Increasing their number by adding the modules supporting popular IoT devices such as: smart TV' audio devices; home automation devices (gates, blinds, heating); sources of light; various types of sensors; IP cameras; the Internet access devices; devices used in gardening; kitchen devices; entertainment systems and car equipment is necessary in order to come onto the market.

Additional modules might also be the modules extending the range of client devices for the service operation. A key change in this area will certainly be the possibility of operation using mobile devices working under iOS, state-of-the-art watches and using voice recognition for giving commands in the system.

A fast growth of both client and end devices may cause that the system in its current version turns out to have reached the end in terms of its performance. If giving commands is taking too much time or adding another device significantly increases a server load, it will not necessarily be the reason for changing the system but only to develop its infrastructure. Owing to the scalability stretched widthwise it is ready to increase its performance considerably before it actually reaches its limits.

The integration of the system with Emotiv EPOC+ NeuroHeadset showed an advanced stage of the development of technologies developing around the BCI. They are no longer only prototypes used in laboratories as it used to be, but devices available for commercial users. Owing to the integration with the system and thus the surrounding devices they become some of the many interfaces of human interaction with equipment, only more futuristic, user-friendly and not requiring mechanical service.

References

1. Paszkiel, S., Matusik, M.: Development of the internet of things in Poland with special consideration of the society's awareness of the IoT. Inform. Technol. Autom. Measur. Econ. Environ. Protect. **2**, 32–35 (2017)
2. Vidal, J.J.: Toward direct brain-computer communication. Annu. Rev. Biophys. Bioeng. **2**, 157–180 (1973)

Chapter 14
Summary

Within the research work carried out, the use in practice of the Moore-Penrose pseudoinverse for the reconstruction of the EEG signal was considered. An essential assumption of the use of the Moore-Penrose pseudoinverse in the EEG signal reconstruction is the possibility of identification of the locations from which a given signal comes. It should be noted that the issue of proper identification of the EEG signal sources is a problematic although not essential issue while constructing the brain-computer interfaces. In this scope, also the use of the LORETA technique, which bases on the concept of solving reverse problems, estimating a distribution of electric activity of neurons in a three-dimensional space was discussed in this monograph. LORETA is a method facilitating visualization of the EEG signals identified as an electric dipole for the purposes of developing a 3D model of the brain. Owing to the LORETA techniques it possible to define more accurately and thus understand the neurological phenomena taking place in human brain based on the information contained in the EEG signal.

As it results from the tests carried out, Matlab environment with EEGLab Toolbox, discussed in this monograph, may be one of the tools for the analysis of the data of the human brain operation in the form of the EEG signal. During the research works conducted, it was also determined that using neural networks it is possible to make generalizations of the EEG data coming from many people in spite of the fact that the brain of each subject of the test participating in the process of learning behaves in a different way. The level of prediction slightly decreases but then many people might use the system implemented based on such a neural network.

The properties of the EEG signal observed during initial analyses might be successfully used for controlling external objects such as, for example, mobile robots. This was successfully used for controlling an object based on Emotiv Inc. software and LabView Emotiv Toolkit V2 package for LabView environment. As it results from the tests performed, controlling a robot in a real time is possible at maximum concentration, but there can always occur some interfering artifacts so an appropriate signal filtration, e.g. using the FFT, is important. A constant improvement and

© Springer Nature Switzerland AG 2020

S. Paszkiel, *Analysis and Classification of EEG Signals for Brain–Computer Interfaces*, Studies in Computational Intelligence 852,

https://doi.org/10.1007/978-3-030-30581-9_14

development of new algorithms for identification and filtration of interference will certainly ensure a wider application of electroencephalographs in automation and robotics. Mobile robots controlled with BCI interfaces may be implemented in the future in many branches of industry and nowadays they are used more and more often in the entertainment sector [1]. Undoubtedly, BCI technology is becoming more common, among others, due to its development by more and more companies and institutions, owing to which a big progress in this field and decrease of the purchase costs of the devices based on this technology can be observed. Owing to the development of BCI technology and robotics, technological achievements created within the above-mentioned technologies are already used, for example, for: controlling wheelchairs by paralyzed people, the disabled and the elderly, the number of whom is growing in present day Europe.

The development of technical sciences and correlation of the conclusions from the scientific research in this field, and a bigger and bigger knowledge of the mechanisms of the human brain operation implies the possibility of developing new solutions that may find a large range of applications. Augmented reality, which originated in the 1960s, is a relatively new discipline which grew out of the technological environment. Brain-computer interfaces, which have been developing dynamically for several years, are a great example of the technology which together with immersive reality may be an interesting tool to be applied in control processes, including mobile robots.

Neurogaming is another interesting application of BCI technology. A positive impact of neurogaming on human body is confirmed in practice by using it for the realization of neuroplastic abilities of the brain within particular needs. Brain operation changes depending on a given person's activity in a given time interval. Therefore, brain fitness, e.g. applications with brain exercises for healthy people, is currently developing dynamically. The development of this field would not be possible if it had not been for a noticeable increase of interest on a global scale and a bigger and bigger number of the devices basing on the brain-computer technology (BCI). In practice, BCI technology is now based on three paradigms: SCP, i.e. slow cortical potentials, P300 evoked potential and ERD/ERS—desynchronization/synchronization connected with a stimulus.

The monograph also presents the solutions which make it possible to connect a few IoT devices and control them using BCI technology.

Reference

1. Ebrahimi, T., Vesin, J.M., Garcia, G.: Brain-computer interface in multimedia communication. IEEE Signal Process. Mag. **20**(1), 14–24 (2003)

Appendix

```
while True:

  syg = getch()

  if syg == ' ':                        #STOP
          GPIO.output(21, GPIO.LOW)
          GPIO.output(20, GPIO.LOW)
          GPIO.output(19, GPIO.LOW)
          GPIO.output(16, GPIO.LOW)
          X5 = 0 # how many % v_ma
          X0 = 115 # as long turn (115 -> 0)

  if syg == 'w':                        #Driving forward
          GPIO.output(20, GPIO.HIGH)
          GPIO.output(21, GPIO.LOW)
          GPIO.output(19, GPIO.HIGH)
          GPIO.output(16, GPIO.LOW)
          X5 = 70 # how many % v_max
          X0 = 115 # as long turn (115 -> 0)

  if syg == 's':                        #Driving backwards
          GPIO.output(21, GPIO.HIGH)
          GPIO.output(20, GPIO.LOW)
          GPIO.output(16, GPIO.HIGH)
          GPIO.output(19, GPIO.LOW)
          X5 = 70 # how many % v_max
          X0 = 115 # as long turn (115 -> 0)

  if syg == 'd':                        #Turn right
          GPIO.output(20, GPIO.HIGH)
          GPIO.output(21, GPIO.LOW)
          GPIO.output(19, GPIO.HIGH)
          GPIO.output(16, GPIO.LOW)
          X5 = 70 # how many % v_max
          X0 = 0 # as long turn (115 -> 0)

  if syg == 'a':                        #Turn left
          GPIO.output(20, GPIO.HIGH)
          GPIO.output(21, GPIO.LOW)
          GPIO.output(19, GPIO.HIGH)
          GPIO.output(16, GPIO.LOW)
          X5 = 70 # how many % v_max
          X0 = 230 # as long turn (115 -> 0)
```

© Springer Nature Switzerland AG 2020

S. Paszkiel, *Analysis and Classification of EEG Signals for Brain–Computer Interfaces*, Studies in Computational Intelligence 852,
https://doi.org/10.1007/978-3-030-30581-9

```python
# calculation v
X5_pwm = sqrt((X5 - 100) * (X5 - 100))
print(X5_pwm)
# Serwo
  serwo = int((X0+180)*2.2)

pwm.setPWM(0, 0, serwo)
pwm.setPWM(1, 0, serwo)
pwm.setPWM(2, 0, int(X5_pwm*20))
pwm.setPWM(3, 0, int(X5_pwm*20))
sleep(0.01)

if syg == 'r':
      print("End")
      sense.clear()
      sleep(1)
      GPIO.cleanup()

      Break
```